3D Printing

by Cameron Coward

A member of Penguin Group (USA) Inc.

This book is dedicated to my lovely wife Sarah, who has always stood by me and supported me. Without her, I could never have done this.

ALPHA BOOKS

Published by Penguin Group (USA) Inc.

Penguin Group (USA) Inc., 375 Hudson Street, New York, New York 10014, USA • Penguin Group (Canada), 90 Eglinton Avenue East, Suite 700, Toronto, Ontario M4P 2Y3, Canada (a division of Pearson Penguin Canada Inc.) • Penguin Books Ltd., 80 Strand, London WC2R 0RL, England • Penguin Ireland, 25 St. Stephen's Green, Dublin 2, Ireland (a division of Penguin Books Ltd.) • Penguin Group (Australia), 250 Camberwell Road, Camberwell, Victoria 3124, Australia (a division of Pearson Australia Group Pty. Ltd.) • Penguin Books India Pvt. Ltd., 11 Community Centre, Panchsheel Park, New Delhi—110 017, India • Penguin Group (NZ), 67 Apollo Drive, Rosedale, North Shore, Auckland 1311, New Zealand (a division of Pearson New Zealand Ltd.) • Penguin Books (South Africa) (Pty.) Ltd., 24 Sturdee Avenue, Rosebank, Johannesburg 2196, South Africa • Penguin Books Ltd., Registered Offices: 80 Strand, London WC2R 0RL, England

Copyright © 2015 by Penguin Group (USA) Inc.

All rights reserved. No part of this book may be reproduced, scanned, or distributed in any printed or electronic form without permission. Please do not participate in or encourage piracy of copyrighted materials in violation of the author's rights. Purchase only authorized editions. No patent liability is assumed with respect to the use of the information contained herein. Although every precaution has been taken in the preparation of this book, the publisher and author assume no responsibility for errors or omissions. Neither is any liability assumed for damages resulting from the use of information contained herein. For information, address Alpha Books, 800 East 96th Street, Indianapolis, IN 46240.

IDIOT'S GUIDES and Design are trademarks of Penguin Group (USA) Inc.

International Standard Book Number: 978-1-61564-744-6
Library of Congress Catalog Card Number: 2014951305

17 16 15   8 7 6 5 4 3 2 1

Interpretation of the printing code: The rightmost number of the first series of numbers is the year of the book's printing; the rightmost number of the second series of numbers is the number of the book's printing. For example, a printing code of 15-1 shows that the first printing occurred in 2015.

Printed in the United States of America

Note: This publication contains the opinions and ideas of its author. It is intended to provide helpful and informative material on the subject matter covered. It is sold with the understanding that the author and publisher are not engaged in rendering professional services in the book. If the reader requires personal assistance or advice, a competent professional should be consulted. The author and publisher specifically disclaim any responsibility for any liability, loss, or risk, personal or otherwise, which is incurred as a consequence, directly or indirectly, of the use and application of any of the contents of this book.

Most Alpha books are available at special quantity discounts for bulk purchases for sales promotions, premiums, fund-raising, or educational use. Special books, or book excerpts, can also be created to fit specific needs. For details, write: Special Markets, Alpha Books, 375 Hudson Street, New York, NY 10014.

Publisher: *Mike Sanders*
Executive Managing Editor: *Billy Fields*
Senior Acquisitions Editor: *Brook Farling*
Development Editor: *Kayla Dugger*
Production Editor: *Jana M. Stefanciosa*

Cover Designer: *Laura Merriman*
Book Designer: *William Thomas*
Indexer: *Brad Herriman*
Layout: *Brian Massey*
Proofreader: *Virginia Vasquez Vought*

Contents

Part 1: What Is 3D Printing? 1

1 The Goal of 3D Printing 3
    For Businesses 4
        Rapid Prototyping 4
        Manufacturing 6
    For the Hobbyist 7
        The Maker Culture 8
        A Tool in Your Arsenal 8
    Common Misconceptions 9
        Materials You Can Use 9
        A Replicator in Every Home 10
        The Time It Takes 11
        What It Can Print 11
        Finishing 11

2 The History and Future of 3D Printing 13
    3D Printing Is Born 13
        Hull's and 3D Systems' Contributions 14
        The Invention of FDM Printing 15
        The Development of Other 3D Printing Processes 15
    The RepRap Project 16
        Initial Intentions 16
        The Importance of Open Source 16
        Rapid Development 17
    Refinement, Availability, and Your Wallet 18
        Maturation of Technology 19
        Availability of Parts 20
        The Race to the Lowest Price 21

3 Manufacturing with 3D Printers and CNC Mills 23
    The Simple Principle Behind 3D Printing 24
        Slicing and Creating Layers for Your Model 24
        Extruding Filament 26

		CNC Milling and How It Works 27
		Subtracting with CAM Software *28*
		Milling the Material *29*
		Pros and Cons of 3D Printers vs. CNC Mills 29
		Cost *29*
		Part Geometry *30*
		Material Matters *32*
		Surface Finish *32*
		Using the Right Tool for the Job 34
	4	**Types of 3D Printers 37**
		Stereolithography 38
		Digital Light Processing 38
		Powder Bed Printing 39
		MultiJet Printing 41
		Selective Laser Sintering 42
		Fused Filament Fabrication 42
Part 2:		**All About the Hardware 47**
	5	**The Frame 49**
		Cartesian Layouts 50
		Ways Cartesian 3D Printers Are Constructed *50*
		Cartesian Layout Considerations *51*
		The Importance of Frame Construction 53
		Rigidity and How It Affects Quality and Reliability *54*
		What Makes a Good Frame *54*
		Size Matters 56
	6	**Movement Components 59**
		Components for Smooth Linear Motion 60
		Rails and Smooth Rods *61*
		Bearings *62*
		Stepper Motors 63
		Belts and Pulleys 65

	Weight-Bearing Components for Converting Motion	67
	Lead Screws	*68*
	Threaded Rods	*69*
	Attachment and Connection Components	70
	Couplers	*70*
	Nuts	*72*
7	**The Extruder**	**75**
	What Is Extrusion?	76
	The Cold End	76
	Direct Feed	*77*
	Bowden	*78*
	Direct Drive vs. Geared	*80*
	The Hot End	81
	Thermistor	*82*
	Heating Element	*83*
	Physical Design and Makeup	*83*
	The Nozzle	85
	Print Fans	86
	Using Multiple Extruders	87
	For Support Material	*87*
	For Filaments with Different Properties	*88*
8	**The Build Platform**	**91**
	Build Platform Materials	92
	Heated Beds	92
	Heated Build Chambers and Enclosures	94
	Surface Treatments	95
	Painter's Tape	*95*
	White Glue	*96*
	Polyimide Film	*96*
	PET Film	*97*
	ABS Juice	*98*
	Hairspray	*98*

9   Control Components 101
   End Stops 102
      Mechanical End Stops *102*
      Optical End Stops *104*
      Other Kinds of End Stops *105*
   Control Boards 106
      Arduinos and Proprietary Control Boards *106*
      Changing or Upgrading Control Boards *108*
   SD Card Support 109
      Using an SD Card *109*
      Benefits of SD Cards *110*
   LCD Controllers 110

10 Choosing a 3D Printer 113
   Open or Closed? 114
      What Does It Mean to Be Open Source? *114*
      Why It Might Matter to You *116*
   Assembled, Kit, or DIY? 116
      Assembled 3D Printers *116*
      3D Printer Kits *117*
      Building a DIY 3D Printer *117*
   Determining Your Needs 118
      Print Volume *119*
      Print Resolution *119*
      Filament *120*
      Prices *121*
   Printers with Unusual Designs 122

Part 3:   Setting Up and Printing 125

11 Software Setup and Printer Control 127
   Firmware Explained 128
   Choosing Host Software 129
   Choosing Slicing Software 131
   Connecting to Your Printer 132
   Controlling Your Printer 133

G-Code 135
    Printing with G-Code *135*
    Performing Functions Manually with G-Code *136*

12   Leveling the Bed and Adjusting the Z Height 139
Why Does Your Bed Need to Be Leveled? 140
    Manual Leveling *140*
    Auto-Leveling *142*
How to Adjust Your Z Height 144
    Setting the Z Height Manually or Automatically *144*
    Knowing the "Correct" Z Height *145*

13   Slicing and Printing 147
Configuring Your Slicing Software 148
Slicer Settings Explained 149
    Printer Settings *150*
    Filament Settings *152*
    Print Settings *154*
Preparing for and Running a Print 157
    Host Preparation *157*
    Running a Print *158*

14   Troubleshooting Your Prints 161
What's the Problem? 161
Extrusion Problems 163
    Overextrusion *163*
    Underextrusion *164*
    Jamming *164*
    Poor Dimensional Accuracy *165*
Temperature Problems 166
    Hot End Is Too Hot *166*
    Hot End Is Too Cold *168*
    Cracking of Part Due to Cooling *168*
    Discoloration of Filament Due to Heat *169*
Adhesion Problems 169
    Warping *170*
    Part Comes Loose *170*
    Part Breaks During Removal *171*

Other Printer Problems 171
    Blobs *171*
    Stringing *171*
    Drooping *172*
    Ghosting *173*

Project 1: Carabiner 175
Preheat the Extruder and Heated Bed 175
Load the .STL File 176
Slice the Model 177
Load the Filament 178
Extrude Some Filament 179
Start the Print 181
Watch the First Layer 181
Let the Part Cool 182
Remove the Part 183

Project 2: Pencil Holder 185
Load the .STL File and Resize the Model, If Necessary 185
Preheat the Extruder and Heated Bed, and Load the Filament 186
Modify the Slicer Settings to Make the Model a Cup 187
Slice the Model 188
Start the Print 189
Watch the First Layer 189
Remove the Part 190

Project 3: Robot 193
Load the .STL File 193
Preheat the Extruder and Heated Bed, and Load the Filament 194
Modify the Slicer Settings for Supports 194
Slice the Model 196
Start the Print 197
Watch the First Layer 197
Remove the Part 198
Remove the Supports 200

Project 4: Storage Box with Drawers..203
- Open Your Host Software and Load the Storage Body .STL File............204
- Slice the Storage Body..205
- Print the Storage Body...206
- Load the Drawer .STL File and Slice It..207
- Print the Drawer..208
- Print Two More Drawers..209

Part 4: 3D Modeling..211

15 Introduction to CAD...213
- Why CAD Came About...214
- Artistic 3D Modeling vs. CAD Software..215
- CAD Software Options..217
- An Overview of Common CAD Program Commands.............................218
  - *Modeling Commands*...*218*
  - *Sketching Commands*..*222*
- The Importance of Units and Scale...223
  - *Choosing Units*..*223*
  - *Scaling in CAD*...*225*

16 Modeling Techniques and Best Practices.......................................227
- Premodeling..228
- Tricks of the Trade...228
- Assemblies and Fitting Parts..231
  - *Planning an Assembly*..*232*
  - *Fitting Parts Together*..*233*
- Modeling Successful 3D Parts..235
- Exporting Files...236

17 Practical Reverse Engineering..239
- Why You Should Learn Basic Reverse Engineering..................................240
- Finding the Right Measurement Tools..240
  - *Digital Calipers: A Necessary Tool*...*241*
  - *3D Scanners: Not All They're Cracked Up to Be*....................................*242*
- Measuring the Part..243

  Inferences 244
   Geometric 244
   Design Intention 245
   Proportions 247
  Modeling the Part 248

Project 5: Monogrammed Coaster 251
  Open Your CAD Program and Create a New Part 251
  Extrude a Circle 252
  Fillet the Top Edge 254
  Cut the Letter 254
  Export the .STL File and Print 256

Project 6: Custom Storage Drawer 259
  Open Your CAD Program and Create a New Part 259
  Extrude the Body of the Drawer 260
  Cut an Opening for the Handle 260
  Design the Compartments 261
  Add a Rough Handle 263
  Add Fillets to Handle 264
  Add Ridges for Grip 264
  Export the .STL File and Print 265

Project 7: Dust Collector 267
  Create a New Part and Revolve the Body 267
  Cut a Groove 268
  Create the Second Part and Revolve the Body 269
  Make the First Clip 270
  Copy the Clip 271
  Export the .STL Files and Print 273

Project 8: Reverse Engineering a Useful Part 275
  Create a New Part and Extrude the Body 276
  Fillet the Edges 276
  Shell the Cover 277
  Cut One Socket Opening 278
  Mirror the Socket Opening 279

		Create the Screw Hole Support	280
		Add the Screw Hole	281
		Export the .STL File and Print	282

Part 5: Advanced Usage and Techniques 283

18 Printing with Other Materials 285

What Materials Are Available? 286
 Nylon 286
 Polycarbonate 286
 Flexible Filament 286
 Wood Filament 287
 PET 288
 HIPS 288

Hardware Needed 288
 All-Metal Hot Ends 289
 Print Fans 289
 Heated Beds and Bed Materials 290

Printing Techniques 291
 Temperature 291
 Speed 292
 Cooling 292
 Layer Thickness 293

19 Modifying Your Printer 295

Adding a Fan Shroud 296
Adding a Heated Bed 297
 Bed Size 297
 Control Board 298
 Power Supply 298

Switching to All-Metal Hot Ends 300
Installing Multiple Extruders 301
Extending Axes 302
Converting to a PCB Mill 302
Alterations for Laser Cutting 303
Finding Parts 304

Appendixes

A	Glossary	307
B	Resources	313
C	Further Uses of 3D Printing	315
	Index	323

Introduction

In many ways, 3D printing is the fulfillment of the future we were all promised. We may not have flying cars or robot housemaids, but we do have access to an almost magical device that can make our imaginations a reality. It wasn't long ago that such a device was considered far-fetched, even by science fiction standards.

I've been a lifelong geek myself. As long as I can remember, I've been building things, taking things apart, and putting them back together (sometimes successfully). When I was very young, this mostly meant building things like spaceships with Legos. When I got a little bit older, I gained access to hammers and nails, and suddenly I could build things at a life-size scale (though this mostly resulted in lopsided tree houses).

Like many geeks my age, I found computers in my teenage years. This opened up a whole new world for me. No longer was I constrained by the cost of wood and nails, because I could create programs on old computers I found at garage sales. Into adulthood, computers were my life. I built them, programmed them, and repaired them.

But the desire to create something tangible never left me. I knew I wanted to make things. So I went to school for drafting and mechanical design. After graduating, I got a job in the engineering team at a biomedical company. While working there, our engineering department got a professional 3D printer for making prototypes of our products.

I was, of course, instantly smitten. Here was a machine that could produce a physical part from my designs in just a few hours. As you might expect, everyone on the engineering team felt the same way. In between prototyping parts, we took turns printing designs for our own personal projects.

But I wasn't satisfied to just print things at work when the printer was available, so I purchased an inexpensive consumer 3D printer. Suddenly, I could print anything I wanted at any time. If some idea struck me, I could design it and print it in just a few hours. It brought back that feeling of building spaceships out of the Legos that I hadn't felt since I was a kid. That feeling of being able to create anything I could imagine.

It's a feeling that most people never find again after childhood. But, if you're reading this, I suspect you might be searching for that feeling yourself. Even if you've just picked this book up at the bookstore out of curiosity, I think that means you're yearning to flex those creative muscles once again.

And that's what 3D printing can do for you. Yes, it's challenging. No, it's not a particularly cheap hobby. But it's a very rewarding one. The feeling of pride you get when your first print is completed will be undeniable. The smile that crosses your face the first time you see one of your own designs made real will be one of the brightest you've had in years. If you're looking for a way to express your creativity, 3D printing is the answer. This book will help you through the complicated task of learning how to do that.

How This Book Is Organized

This book has been written to help you really understand 3D printing, not just to teach you the basics. To that end, the book has been split up into five parts:

Part 1, What Is 3D Printing?, gives you useful information on the background of 3D printing and how it actually works. This includes the history of 3D printing, how 3D printers developed into a consumer product, and how different kinds of 3D printing processes work.

Part 2, All About the Hardware, teaches you everything you need to know about the different parts of consumer 3D printers. Learning how the hardware of a 3D printer actually works is valuable knowledge that can help you a lot throughout your journey. (And, because you're reading this book, I suspect you're the type of person who likes to learn anyway.)

Part 3, Setting Up and Printing, is where you start getting into the nitty-gritty of 3D printing. This is where you learn how to actually set up and operate a 3D printer. 3D printers aren't like the old inkjet printer you have on your desk; they're complicated machines that are chock full of confusing settings. This part gives you the information you need to really get started. I also walk you through four projects designed to teach you how to get started with 3D printing.

Part 4, 3D Modeling, goes beyond just 3D printing models you find online, because 3D printing is about so much more than just replicating other people's designs. Can you imagine if, as a kid, you were only allowed to build official designs with your Legos? It would have been awful! For that reason, this is where you learn how to design your own 3D models that you create with a 3D printer. In this part, I also show you how to start 3D modeling with four more projects for you to model and print.

Part 5, Advanced Usage and Techniques, is what you want to read once you've gained proficiency in basic 3D printing. Here, you learn how to print with exotic materials, how to print with multiple extruders at once, and some popular modifications you can make to your 3D printer. You also learn about some unexpected uses for your 3D printer that you may not have even realized were possible.

Extras

Throughout this book, you'll find some handy bits of information and advice beyond the main text in sidebars. They're broken up into a handful of types for your convenience.

DEFINITION

The 3D printing industry is full of technical terms and words that may have a meaning specific to 3D printing. These sidebars include words you're not familiar with, or that have a different meaning in the context of 3D printing.

WATCH OUT!

While 3D printers are now consumer products that are safe for the home, they're still machines that can hurt you. Therefore, keep an eye out for these warning sidebars, which give you safety information and help you avoid accidentally damaging your 3D printer.

HOT TIP

Sometimes, there are multiple ways of doing something. These tips help you avoid hassles and headaches. Learn from my mistakes, and take these into account!

FASCINATING FACT

I can't always fit the good stuff into the main body of the text. Check these out for interesting facts, history, and other information related to 3D printing.

But Wait! There's More!

Have you logged on to idiotsguides.com lately? If you haven't, go there now! As a bonus to the book, we've included the 3D printing model files you'll need for the Part 3 projects, all online. Point your browser to idiotsguides.com/3dprinting, and enjoy!

Acknowledgments

I could never have written this book without the support of my friends and family, who have always encouraged me. My beautiful wife Sarah, who put up with me working late into the night. My friends Daniel, Sean, Sam, Zach, Amanda, and Jonathan, who were always happy to start new hobbies with me. My in-laws Bonnie and Javad, who somehow believed in me when no one should have. My parents Sharon and David and siblings Danielle, Lynnze, and Nathin, who had to live with me for so long. Tom, Beth, and Mike, who nurtured my inner geek when I was young. And my dog Calaveras, who kept me company while I wrote this book, and because she's just so darn cute.

I'd also like to give a special thanks to the fine folks at Aleph Objects (especially Harris Kenny and Jeff Moe), who were kind enough to lend me a LulzBot TAZ to use for this book. The RepRap project also deserves a huge amount of credit for making 3D printing possible for hobbyists and consumers.

Special Thanks to the Technical Reviewer

Idiot's Guides: 3D Printing was reviewed by an expert who double-checked the accuracy of what you'll learn here, to help us ensure this book gives you everything you need to know about 3D printing. Special thanks are extended to Aaron Trocola.

Trademarks

All terms mentioned in this book that are known to be or are suspected of being trademarks or service marks have been appropriately capitalized. Alpha Books and Penguin Group (USA) Inc. cannot attest to the accuracy of this information. Use of a term in this book should not be regarded as affecting the validity of any trademark or service mark.

PART

1

What Is 3D Printing?

Every topic needs some background information, and 3D printing is no different. In this part, you learn about the history of 3D printing, including how it was invented and how it entered the consumer space. You also learn about the different kinds of 3D printing technologies available, and the pros and cons of each.

This part also covers the principles behind how 3D printing actually works. In addition to how it works, I go over how it differs from other prototyping and manufacturing methods. This includes how 3D printing can be used by businesses, and what you can use it for personally.

CHAPTER 1

The Goal of 3D Printing

It wasn't that long ago that 3D printing was a just an obscure technology relegated to prototyping use by a handful of specific industries. But in the past couple of decades, 3D printing has become a must-have technology for engineering companies, and even more recently a useful consumer tool.

But why did 3D printing become such an indispensable tool for businesses? What advantages do 3D printers give the consumer and hobbyist? In this chapter, we look at why 3D printing is becoming so popular, what it can do for you, and what its limitations are.

In This Chapter

- Why businesses invest in 3D printers
- What you can do with 3D printing
- Common misconceptions about 3D printing

For Businesses

I say this completely without hyperbole: 3D printing is a dream for an engineering team. A technology that would have been the stuff of fantasy just a few decades ago, today it gives companies the ability to bring an idea to life quickly, efficiently, easily, and inexpensively.

The original intent for 3D printing was to create prototypes for engineers. This allowed them to quickly test an idea before moving on to production and manufacturing. Very recently, some companies have even begun to use 3D printing for the actual manufacturing of production parts. Both uses have exploded in popularity, and both benefit businesses in a number of ways.

Rapid Prototyping

Rapid prototyping, simply enough, is any process that can be used to quickly create a prototype of a part. A prototype part is essentially a test part, and is useful during the development of a new product (or when revising an existing product). What the prototype is intended to test is up to the engineer, but it's commonly used to test the following:

- How parts fit together in an assembly
- The strength of the part
- Regulation compliance

When a new product is being developed, it's rare that a single design is kept until production. Commonly, a handful of potential designs are tested, and whichever design is chosen for production will evolve quite a bit throughout the design process. Much of the testing of these designs is done via simulation with *computer-aided design (CAD) software*, but it's almost always necessary to have a real physical part to work with during this process.

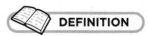
DEFINITION

Computer-aided design (CAD) software is a type of computer program used by engineers and designers to develop products using 3D models or 2D drawings. Before CAD software, drawings were hand drawn by drafters for production. Now, CAD software can export models directly to 3D printers (see Chapter 15 for more on CAD).

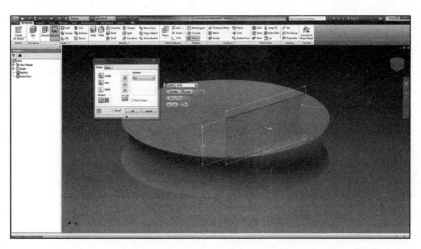

CAD software being used to create a 3D model.

That's where rapid prototyping, and especially 3D printing, comes in. As the design of a product evolves, engineers can 3D print the design and test it. In the past, prototypes were costly and often required outsourcing to companies who specialized in producing these prototypes using traditional methods like machining, welding, sand casting, and other costly processes. Now, it's commonplace for an engineering team to have a 3D printer in their office with which they can print a prototype in a matter of hours and for a relatively low cost.

Aside from the physical testing of the part, prototypes are also valuable for user testing, design proposals, and promotion. When a new product is in the early stages of development, it doesn't always have funding secured for the entire design process. Project funding (whether internal to the company or from outside investors) often depends on a design submittal, which is essentially a proposal of the product design.

Design proposals can be anything from a simple explanation of the idea to a complete working prototype. But 3D printing has made it possible to speed up this process and has made it much cheaper. Therefore, physical prototypes have recently become a virtual necessity for securing outside funding via crowdsourcing. Backers of crowdsourcing want to see that a design is functional before they invest, and 3D printed prototypes are the cheapest way to show that.

Manufacturing

While rapid prototyping is definitely the most common use of 3D printing, it's starting to become a viable method of actual production manufacturing. Traditionally, injection molding is by far the most common method of manufacturing plastic parts. In mass production, injection molding can produce high-quality parts very quickly and inexpensively.

Injection molding is a fairly simple process in concept: molten plastic is shot into the cavity of a two-piece (or more) metal mold and cooled, and then the mold is pulled apart and the plastic part is removed. In a manufacturing setting, that process can happen very quickly and a new part can be made in a few seconds. It can also be very inexpensive (per part), as the plastic itself is cheap in raw form.

However, injection molding has one critical downside: the cost of tooling. Tools, in reference to injection molding, are the molds and associated part-specific machinery needed to produce a particular part. Every individual part needs its own tooling, and that can be very expensive. Molds are usually made from steel or aluminum and require precision machining. Mold making is a specialized task, and individual molds start in the thousands of dollars and often reach into the tens of thousands of dollars.

When you take into account the fact that most products are assemblies made up of many individual parts (which would each need their own tooling), it's obvious that the costs of getting set up for injection molding can be very high. That's okay if you're making thousands of that part, because the cost per part gets lower as the total quantity increases. But what if you want to produce your parts in smaller quantities? The parts must be manufactured in huge quantities in order for the price per part to reasonable.

FASCINATING FACT

Aleph Objects, a 3D printer manufacturer based in Loveland, Colorado, uses its own 3D printers to manufacture some of the parts to build new printers. This "cluster" of 3D printers used for manufacturing is the largest in the world, consisting of 135 individual 3D printers.

That's where 3D printing has started to enter the manufacturing world. With 3D printing, there is no high tooling cost up front—your printers can produce whichever part you need them to at a given time. You can print one part or a hundred, and your price per part will be the same. This means you can manufacture in low quantities without incurring huge tooling costs. Possibly even more valuable is the ability to change designs on the fly. You can redesign a part and immediately start manufacturing the new design without having to purchase a new mold, which means your product can evolve in real time as you improve the design.

The cluster of 135 3D printers used for manufacturing at Aleph Objects.
(Photo courtesy of Aleph Objects)

Of course, 3D printing in a manufacturing setting does have its downsides, which have kept it from gaining widespread adoption. If a part is going to be manufactured in large quantities, it's still cheaper to have it injection molded (despite the cost of the molds). That's because the raw material is much cheaper for injection molding than it is for 3D printing. Possibly even more importantly in mass production is time—an injection-molded part might only take a few seconds to make, while a 3D-printed equivalent can easily take hours.

To be able to keep up with an injection molding machine, many 3D printers would have to be run simultaneously. So injection molding is still the best option when parts need to be produced in mass quantities. But for small quantities or designs which need to be updated often, 3D printing is becoming an attractive alternative.

For the Hobbyist

Engineering 3D printing prototypes for corporations is all well and good, but what's really exciting is 3D printing for the average person at home! In just a couple of decades, 3D printers have gone from being too expensive for most companies to buy to being cheap enough that you can realistically purchase one just to tinker with at home. But why would you want one in your home? What could you even do with it?

The Maker Culture

The idea of making stuff for fun and self-fulfillment has gained tremendous popularity recently, but the basic motivation is an inseparable part of human nature. The same urge to invent that put man on the moon is also what motivated your father to build that birdhouse when you were a kid, and it's the same thing that has you interested in 3D printing. We all have this urge to create (to varying degrees, of course), and the maker culture is just a modern branding of that—a way to identify like-minded people and form a community.

3D printing has been a huge factor in the growing maker community, and for good reason. Traditional maker activities include things like wood working and metal working. Those are skills which take a great deal of practice, patience, time, and money. But 3D printing allows you to create similar physical objects quickly and relatively easily. It's become a fairly simple task to take the ideas from your head and make them real.

At its heart, maker culture is about creating things. It's using whatever tools you have available creatively to invent and construct things that interest you. In the same way that your father may have used a hand saw, a hammer, and nails to build that birdhouse, you can use a 3D printer to make whatever you can imagine. Robots, electronics enclosures, and artistic creations are all possible, along with anything else that interests you.

A Tool in Your Arsenal

By definition, a 3D printer is just a tool, and that's how you should think of it. If you want to hammer nails, you use a hammer. If you want to cut a piece of wood, you use a saw. And if you want to make a part out of plastic, you use a 3D printer. It's important to think of a 3D printer like any other tool—a very versatile tool, but still a tool.

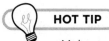
HOT TIP

Make sure to utilize 3D printing in the right context. It's impractical to try to 3D print an entire coffee table, but you can certainly use your 3D printer to make brackets, drawer handles, tracks, and other parts. Like any other tool, a 3D printer has uses for which it's well suited, and some things which would be better handled by another tool.

As always, it's best to use the right tool for the job. It might be possible to cut a metal pipe with a wood saw, but it's not a good idea. 3D printers are no exception—they can do a lot, but they're not the right tool for every job. When you first start 3D printing, you'll be tempted to try and use it for everything. Birdhouse? 3D print it! A coffee table? Print the pieces and glue them together! But you'll quickly learn that 3D printers aren't the solution to every problem.

With that said, 3D printers are incredibly versatile. That's why they're such a revolutionary tool. You can use them to print everything from things like pencil holders to replacement parts for your car. This is what makes 3D printing so exciting: if it fits on the printer and it's made of plastic, you can make it! Sure, 3D printers aren't the solution to every problem, but they're one of the most widely useful tools ever invented.

Common Misconceptions

In essence, the goal of 3D printing (whether for business or consumer use) is to be able to quickly and easily create objects with a single device. To an extent, 3D printers today do accomplish that feat.

But 3D printing presents a kind of perfect storm of misunderstanding that many fall victim to. It's a technology that is very new to consumers, it's complex, and it's still fairly experimental. It's somewhat reminiscent of the early days of the internet, when most people still didn't quite know what to make of it. There are some real-world drawbacks which keep them from being the miracle devices envisioned in popular science fiction and media.

Materials You Can Use

Probably the most common of these misconceptions concerns materials. 3D printing technology has come a long way, and there are a variety of materials available to print with. But, for consumer printers at least, all of these materials are still some type of plastic. Printing metal is out of the question for a consumer printer. Metals, ceramics, and other more exotic materials are possible on extremely expensive professional 3D printers, but don't expect to be printing them at home anytime soon.

This is primarily because the technology needed to print in metal or ceramics is far more complex than what is needed for a consumer 3D printer. Consumer 3D printers use fused deposition modeling (FDM) or fused filament fabrication (FFF) technology, which is inexpensive (see Chapter 2). But to print in metal, for example, an expensive and complex process like selective laser sintering (SLS) is required. SLS 3D printers use high-powered lasers to actually melt metal powder, and, as you can imagine, those aren't cheap. So for now, consumer printers are relegated to just printing plastic.

A Replicator in Every Home

A popular theme in news articles about 3D printing is the idea of a replicator in every home. It's the picturesque future from the popular culture of the '60s and '70s: a device that instantly produces anything you need at the touch of a button.

FASCINATING FACT

The replicator was first introduced in *Star Trek: The Original Series* (and officially named in *Star Trek: The Next Generation*) as a device that could recycle garbage and waste and synthesize food and other useful things. The idea is so similar to 3D printing that a popular line of 3D printers made by MakerBot even uses the replicator name for some of their models.

The setup is simple: there you are in your kitchen cooking dinner, when suddenly your spatula breaks. Your dinner is starting to burn, and you don't have time to drive up to the store and buy a new spatula. Not to worry though, you have your trusty 3D printer! You run over, push a button or two, and in a matter of minutes you have a new spatula. Problem solved! This is the kind of thing imagined by the media and those unfamiliar with the realities of 3D printing.

Unfortunately, reality doesn't quite meet the expectations set by these kinds of stories. 3D printer developers would certainly love to be able to produce such a product, but it's just not the case (and probably won't be for a very long time).

Your dinner fiasco would be quite a bit different in real life today: your spatula breaks, and dinner is about to start burning. You turn to your 3D printer for help. But before you can start printing that new spatula, you've first got to find a 3D file to print from. After searching for a few minutes online, you're lucky and find one (saving yourself from having to 3D model one). Now all you have to do is load the file, heat up the printer, start it printing, and wait an hour or two for it to print. Of course, by that point, your dinner is already burned.

As you can see, as amazing and versatile as 3D printers are, they do have their limits (of which there are many). Eventually, you may have that sci-fi replicator in your house, but we're not quite there yet.

The Time It Takes

The time involved to print something is also a highly misunderstood topic. 3D printing is often referred to as being very fast (I've even said so earlier in this chapter). But it's important to note that that's only in relation to traditional manufacturing methods. Even the smallest prints are going to take a few minutes at best. Larger prints on consumer 3D printers can take the better part of a day, and it's possible for a single print to take days (even on an average-size consumer printer). So don't expect to be printing out parts instantly; it's usually a fairly long wait.

WATCH OUT!

3D printers have very hot parts and can be a significant fire hazard. Only leave a 3D printer unattended at your own risk, and be sure to follow appropriate fire safety precautions when printing with it.

What It Can Print

3D printers carry with them the promise of being able to print anything. But there are some caveats when it comes to what geometry you can actually print. Overhangs are the biggest enemy here. Overhangs are geometry which have no supporting material directly underneath, which is troublesome for obvious reasons. This can be overcome with support material, which I'll go over in more depth in later chapters, but you should be aware of the fact that 3D printing does have design challenges of its own.

Another area where 3D printers (especially consumer 3D printers) have difficulty is with printing small features with fine details. They're limited by the size of the filament being extruded (or pushed out) from the nozzle, and fine detail tends to get washed out when common nozzle sizes are used. Professional 3D printers that use more expensive technology are capable of astounding detail, but consumer 3D printers are still pretty limited.

Finishing

The finish of the printed object is one of the most glaring limitations of 3D printing. Because of the nature of the process (essentially stacking layers of plastic), the surface finish of vertical surfaces is usually quite poor. Even on the highest-quality settings, you'll still be left with a series of small ridges. It's virtually impossible to get a truly smooth finish straight from a consumer 3D printer. People expend a great deal of effort trying to minimize this and improve surface finish, and there are some postprint techniques you can use to smooth it out. But don't expect to get a nice, smooth part hot off the printer.

Even with high-quality settings, layer ridges are evident on this owl model.

You shouldn't be too disheartened by all of this, though. These are all fairly minor limitations, and are just part of 3D printing. There are ways to deal with most of them, and a lot of work is being done to overcome these challenges. It's important to keep your expectations for 3D printing realistic, but don't let these downsides trouble you too much.

The Least You Need to Know

- 3D printers initially gained popularity for rapid prototyping, which remains their most common use today.
- Manufacturing with 3D printers is a viable option for small-scale production but is still impractical for mass production.
- The maker culture has been fueled by 3D printing and the creative outlet it provides.
- Being able to produce anything in your own home is an exciting idea, but it's far from being a reality.

CHAPTER

2

The History and Future of 3D Printing

If you want to understand something, it's always a good idea to begin by looking at the history. It'd be difficult to truly understand how a car works, for example, if you didn't know how the internal combustion engine was developed. 3D printers are no exception to the rule, and for such a new product they have quite a rich history.

In this chapter, I take you through the history of 3D printing and the effects of its evolution on you now and in the future.

In This Chapter

- The development and history of 3D printing
- How the RepRap project accelerated development
- Predicting the near future of consumer 3D printers

3D Printing Is Born

3D printing has just entered the public consciousness in the past few years as consumer 3D printers have become available. But 3D printing has actually been around for decades in the industrial world. 3D printing revolutionized the way companies produce prototypes, and that provided the basis for consumer 3D printing.

So how was 3D printing actually developed? Who invented it and why?

Hull's and 3D Systems' Contributions

The first patent for a 3D printer was filed in 1984 by Charles W. Hull. His patent was for a 3D printing process called *stereolithography (SLA)*, which used UV light to cure *photopolymer resin* in a vat to form parts. Hull, an engineer specializing in materials science at the time, came up with the idea for SLA while working on resin coatings for tabletops.

> **Photopolymer resin** is a type of liquid resin that solidifies into plastic when exposed to light (usually in the ultraviolet spectrum). Manufacturers can produce the resin in many varieties, with different mechanical and chemical properties.

The company Hull was working for at the time, Ultra-Violet Products (UVP), employed him to develop UV-curable coatings to improve the durability of tabletops. These coatings were a liquid resin that reacted with UV light to become solid plastic. While working with this resin, Hull started imagining a device which could cure this resin in successive layers to form a three-dimensional object.

Hull took his idea to UVP and was granted permission to work on a prototype device on nights and weekends, while continuing his normal duties during the day. The development process was not without its hurdles, one of the biggest being how to translate a 3D computer model into printing instructions for his SLA printer.

At the time, in the early '80s, computer-aided design (CAD) was still in its infancy. 3D modeling was a difficult process, and not a whole lot could be done with the resulting models. Hull knew he needed 3D models in a file format that could be used to create instructions for his printer, but no suitable file format existed. So Hull created his own: the SLA file format, abbreviated as STL, which is still the standard today. (These days, because the STL file format is used in a wide range of manufacturing processes, STL is often considered an abbreviation for *standard tessellation language*.)

The STL file is created by taking a 3D model from CAD software and converting it into a surface mesh consisting of many triangles. The beauty of the STL format is that the number of triangles determines the detail of the resulting surface mesh, making it scalable. With this file format in hand, Hull was able to create software to translate the 3D model into a series of instructions for his printer to follow.

In 1983, Hull successfully printed his first 3D model: a basic cup. Knowing he had a viable and useful new method of rapidly creating prototypes, he filed his patent for SLA in 1984. In 1986, his patent was granted, and in the same year he created his company 3D Systems to develop and sell SLA printers.

FASCINATING FACT

While Charles W. Hull did patent his SLA process, he didn't patent the STL file format that he developed. This has allowed STL files to be used by other 3D printer manufacturers and even in other types of machines, like computer numerical control (CNC) mills. For this reason, the STL file format has become the standard for 3D printing and CNC milling.

The Invention of FDM Printing

The fused deposition modeling (FDM) printing process—a technology used by the vast majority of consumer printers today—was originally developed by Scott Crump in 1989, in order to ease the process of prototyping at IDEA, Inc., a company he cofounded in 1982. Crump, along with FDM printing itself, also developed some of the necessary associated technologies (such as ABS filament, which I'll discuss more in Chapter 4).

After inventing the FDM 3D printing process in 1989, Crump founded Stratasys—currently the largest 3D printer manufacturer—with his wife Lisa. In 2009, Stratasys's FDM printing patent expired, opening up the market for consumer FDM 3D printers, usually referred to as *fused filament fabrication (FFF)* for non-Stratasys 3D printers.

The Development of Other 3D Printing Processes

During the same time frame that Stratasys and 3D Systems were developing SLA and FDM 3D printing, other 3D printing processes were being developed independently. At the University of Austin in the mid-1980s, selective laser sintering (SLS) was developed by Dr. Carl Deckard and Dr. Joe Beaman with sponsorship from Defense Advanced Research Projects Agency (DARPA). This technology was originally sold by DTM Corporation, which was then purchased by 3D Systems in 2001.

Meanwhile, in the early '90s at the Massachusetts Institute of Technology (MIT), inkjet 3D printing was invented. Z Corp. gained the license for this technology and produced inkjet 3D printers until 2012. On January 3, 2012, Z Corp. was purchased by 3D Systems in order to acquire the associated inkjet printing patents and licenses.

Due to the expiration of key patents, many of these technologies are starting to enter the consumer market (or will be soon). But it was FDM printing that jump-started consumer 3D printing. This was mostly thanks to the RepRap project.

The RepRap Project

3D Systems, Stratasys, and others revolutionized research and development with 3D printing. But for many years, 3D printers remained expensive and complex tools. The high learning curve to use them meant that most users needed special training, and their high cost meant they were out of reach for an individual person. That remained true until the RepRap project was launched in 2005 by Dr. Adrian Bowyer, an engineering lecturer at a university in the United Kingdom.

Initial Intentions

When Dr. Bowyer first founded the RepRap project, his intentions were simple: to develop an inexpensive open-source 3D printer, with a long-term goal of self-replication. The name "RepRap" is a contraction of "replicating rapid prototyper." The idea is that the best way to get a 3D printer into as many hands as possible is to design a 3D printer that can 3D print a copy of itself. If each person prints two new printers for friends, the spread would be exponential, and before long, everyone would have small-scale manufacturing at their fingertips.

The goal of self-replication is a long way off from fruition; it is a lofty goal, after all. Such a design would have to be capable of not only printing the plastic frame pieces, but also motors, electronics, and other complex nonplastic parts. But that optimistic long-term goal has yielded bountiful developments for consumer 3D printers.

The Importance of Open Source

One of the core tenants of the RepRap project, possibly the most important of them, is that it is completely open source. Every design and development is completely public, allowing anyone to use the designs and contribute to the project.

Anytime a new breakthrough is made, it can be immediately released and integrated into the design of new printers. A good example of this is the RepRap Arduino Mega Pololu Shield (RAMPS) control board, which is essentially the brain of the 3D printer. The RAMPS board connects to an *Arduino* (another open-source project) to provide computer control of the 3D printer's motors and extruder. The development of RAMPS allowed people to quickly, cheaply, and easily control their 3D printers, so it was quickly integrated into the design of subsequent RepRap printers.

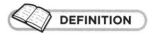

DEFINITION

Arduino is an open-source platform for developing and prototyping electronics. Arduino models are generally small circuit boards without inputs and outputs for controlling various electronics. The Arduino Mega is used by the RAMPS control board for 3D printing.

The same is true of any other new RepRap development. Because it's open source, there are no patents to deal with, and the new development can be freely used and improved. This concept has been essential to the success and rapid evolution of RepRap 3D printers, and consumer printers in general.

Rapid Development

Due to their open-source nature, RepRap printers have evolved extremely quickly. The first complete design, the Darwin 3D printer, was released in 2007. Darwin was a very basic design, wrought with limitations and capable of only mediocre-quality printing, but it proved that 3D printers could be inexpensive and built at home by hobbyists. Like most subsequent RepRap printers, and consumer printers in general, Darwin utilized the fused filament fabrication (FFF) 3D printing process. FFF is exactly the same as FDM in practice (filament is melted and squeezed out of a nozzle in lines, forming layers) and was only named differently to avoid legal problems with Stratasys, who patented the FDM process.

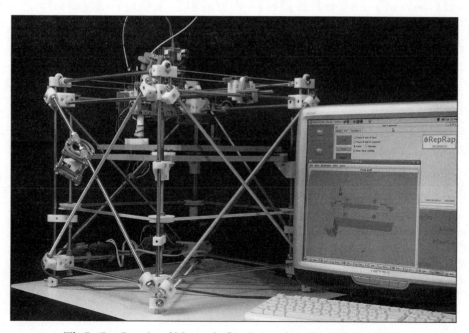

The RepRap Darwin, which was the first design released by the RepRap project.

In 2009, a new RepRap design called *Mendel* was released. It made a number of improvements to the Darwin design, which increased reliability and print quality and decreased the difficulty of building the printer. Mendel integrated many developments that had been made between 2007 and 2009 into a complete package, which would be a continuing trend with new designs that followed.

By the time the Prusa Mendel and Huxley RepRap designs were released in 2010, the community had already grown tremendously. Many offshoots and derivatives of the standard models already existed, making improvements and customizations to suit individual tastes. Today, there are at least 30 official RepRap designs, with derivatives numbering in the hundreds. Current designs rival the consumer printers being released by companies like 3D Systems and Stratasys, who independently developed their own (closed-source) designs. RepRap printers have achieved quality and reliability so high that they're even being used in professional prototyping settings.

FASCINATING FACT

RepRap models, following the convention set by the original Darwin, are named after notable evolutionary biologists. However, some derivatives, such as the Prusa Mendel, are named after their designers or names they chose.

Refinement, Availability, and Your Wallet

The jump from expensive professional 3D printers to consumer 3D printers happened astonishingly quickly. In just a few years, consumer 3D printing went from being experimental, unreliable, and difficult to something that could feasibly be taken advantage of by the average hobbyist. So how did that rapid evolution happen, and what drove it? That was largely due to the development pioneered by the RepRap project.

Until the RepRap Darwin proved that consumer 3D printers were feasible, virtually all 3D printers were expensive professional machines meant for corporate buyers. But with the success of the RepRap project, consumer 3D printing was suddenly on everyone's mind.

3D printer manufacturers specializing in inexpensive models intended for home use began popping up rapidly. Some of these manufacturers sold printers based on RepRap designs, while others developed their own. However, even those that developed their own did so with knowledge and information that often originated with the RepRap project.

In the short time since Darwin was released in 2007, the consumer 3D printer market has grown exponentially. Prices have dropped from the tens of thousands of dollars to just a few hundred dollars for the least expensive printers. But it's not just the price that has changed. The quality of available 3D printers has evolved by leaps and bounds as well. Just like with the Cambrian Explosion, the market has gone from being essentially nonexistent to being a huge and diverse ecosystem full of evolved and refined 3D printers.

What does that explosion of 3D printer evolution in recent years mean for the current market? With such a short history, you might expect that 3D printers aren't quite ready for mainstream use, and to an extent you're right. 3D printing in the home is still pretty experimental, and it's not as simple as printing on a sheet of paper with an inkjet printer. But 3D printing is a revolutionary and exciting technology, and the pace at which it's improving is astounding.

Maturation of Technology

No matter what the particular product is, you can bet that it will improve as the associated technology matures. This has been especially evident since the advent of personal computing. The first personal computers were big, expensive, slow, and weren't particularly useful. But the technology improved so quickly that within a few years, prices had dropped dramatically and usefulness had improved exponentially. Of course, now the technology is so good that computers are ubiquitous.

A similar metamorphosis is happening with 3D printers right now. We're already past the "expensive and useless" stage now, and we're well into 3D printing entering mainstream use. As the technology has matured, proven formulas and design principles have started to emerge.

In the early days of consumer 3D printing, everything was still highly experimental. Engineers were still trying to figure out the best way to design printers, individual parts were still being cobbled together from whatever was available on the market, and printing methods and practices were still being developed. Now we've entered a stage of refinement. We've got the basics down, there is a solid knowledge base with which to work from, and industry standards have started to form.

At this point, printer manufacturers are no longer trying to simply invent a practical 3D printer; instead, they're improving the functionality of proven designs. In the same way that auto manufacturers refine existing engine designs rather than inventing new engines, 3D printer manufacturers are now working on refining their printers to make them better and more useful.

Availability of Parts

As a technology matures, there is also a significant increase in the availability of parts specific to that industry. When a company like Dell decides to build a new laptop, they don't manufacture all of the individual parts from scratch. Instead, they purchase processors from Intel, memory from Corsair, hard drives from Seagate, and so on. These parts are all standardized, and Dell can purchase them knowing they'll work together. They can then assemble the parts into a working computer of their design.

FASCINATING FACT

Just 10 years ago, in 2004, the most inexpensive 3D printers available still cost about $25,000. Now, in 2014, functional 3D printers can be purchased for as little as $200. This perfectly illustrates the effect that development and parts availability has had on 3D printer prices.

The same process is true in most industries. But until recently, 3D printing was so new that individual parts simply weren't available to printer manufacturers. Manufacturers were forced to either repurpose parts used in other industries (which may not be well suited to 3D printing) or make them themselves (in which case, quality and cost became a problem). The lack of parts available in the 3D printing industry was a huge factor in the high costs of early consumer printers.

But that has started to change. Many parts are starting to become available to printer manufacturers. Just like Dell can order a processor from Intel, a 3D printer manufacturer can now order a hot end, control board, or heated bed for use in their printers. This has greatly reduced the costs involved in designing and manufacturing 3D printers and has also improved quality.

Instead of the printer manufacturer having to try and design each and every individual part with marginal results, they can now order them from a company which specializes in that individual part. E3D, for example, is a company that makes hot ends for 3D printers, and that's all they do. Because it's their sole focus, E3D can concentrate on just trying to make the best hot ends on the market. Printer manufacturers can then purchase those hot ends for their printers and have a high-quality part without a massive design and manufacturing investment.

Chapter 2: The History and Future of 3D Printing

A hot end manufactured by E3D that's intended for use on a variety of printers.

With companies like E3D producing parts specifically for the 3D printing industry, the quality of consumer 3D printers has increased very noticeably. And that will only get better as new manufacturers of parts enter the market and competition increases.

The Race to the Lowest Price

That competition between part manufacturers, and the competition between 3D printer manufacturers, has done wonders for the consumer market. Competition in the marketplace always fuels progress and innovation, and that has been particularly evident with 3D printers because it's happened so fast. In the short time since the RepRap project was started, a large number of 3D printer manufacturers have sprung up. Virtually all of them have the same goal: to provide a high-quality and easy-to-use printer to consumers at the lowest price possible.

Originally, that lowest price was still pretty high. But as parts availability has increased and designs have matured, those prices have been falling like a rock. 3D printers are now under $350, when originally the cheapest consumer printers were still thousands of dollars.

 **HOT TIP**

While prices are certainly going to keep going down, they're already starting to level out. So if you've been waiting to purchase a 3D printer until prices drop, don't expect them to get dramatically lower (at least not for a worthwhile model).

However, the race to have the lowest prices can only go so far before it levels out. Materials and manufacturing are always going to be an unavoidable expense for 3D printer manufacturers, and there will be a point when it's just not possible to sell printers any cheaper. We may not even be that far from that point today.

The good news is that just because prices can't drop anymore, that doesn't mean progress can't still be made. Once prices level out, the new race will be toward making the best printer at that price. A $500 computer that you buy today will vastly outperform a $500 computer from 10 years ago. And something similar is likely to happen with 3D printers in the future. Prices might level out at around a couple hundred dollars, but what you get for that money will continue to get better and better as time passes.

The Least You Need to Know

- 3D printing was originally invented by Charles W. Hull. He also started 3D Systems, which remains one of the largest 3D printer manufacturers. FDM 3D printing—the technology used on most consumer 3D printers—was invented by Scott Crump, who went on to found Stratasys with his wife Lisa.
- The consumer 3D printer market was largely nonexistent until the RepRap project fueled open-source development.
- 3D printer prices have decreased dramatically as the technology has matured and parts have become available. Quality, reliability, and ease of use have simultaneously improved as a result.
- While the cost of 3D printers is sure to lower a little bit more, prices are starting to level out. However, the printers themselves will almost definitely improve a great deal in coming years.

CHAPTER 3

Manufacturing with 3D Printers and CNC Mills

When it comes to rapid prototyping, there are generally two approaches that can be taken: 3D printing and computer numerical control (CNC) milling. CNC milling isn't strictly considered "rapid prototyping," but that's mostly a matter of semantics. In practice, CNC mills can create 3D parts just like a 3D printer, and can often do it more quickly and with greater precision.

However, CNC mills and 3D printers differ greatly in their actual operation. Essentially, they're two completely different ways of solving the same problem: how to quickly create a functional three-dimensional part from a computer model. 3D printers solve this problem with additive manufacturing, meaning the 3D printer starts with nothing and adds material to create the part. On the other hand, CNC mills use subtractive manufacturing, a process which starts with a block of material and cuts it away, leaving the part behind.

Both processes have their advantages and disadvantages and are useful in different situations. You've purchased this book, so you've probably already decided that 3D printing is right

In This Chapter

- How 3D printers work
- The workings of CNC mills
- Should you use a 3D printer or a CNC mill?

for you. So why even discuss CNC milling? For one, highlighting the differences will help you understand how 3D printing works and its benefits. The second reason is more practical: it's often possible to convert CNC mills into 3D printers, and some manufacturers are even building multipurpose machines that are capable of both CNC milling and 3D printing. In this chapter, I take you through the functions of both 3D printers and CNC mills and how they can be useful to you in the 3D printing process.

The Simple Principle Behind 3D Printing

3D printing is an additive manufacturing process, and how a 3D printer operates is all about laying down new material as precisely as possible. For virtually all 3D printers today, this is accomplished by building the part up in a series of horizontal layers stacked on top of each other. In consumer 3D printing, those layers are almost always created by extruding molten plastic in a precise 2D cross-section of the part.

Slicing and Creating Layers for Your Model

The one thing that virtually all 3D printers have in common, regardless of the particular technology they're using, is that they all build parts layer by layer. Part of the 3D printing work flow is taking the desired 3D model (generally in STL format) and feeding it into slicing software. This appropriately named software then slices that 3D model into a series of thin horizontal cross-sections that will form the complete part when stacked.

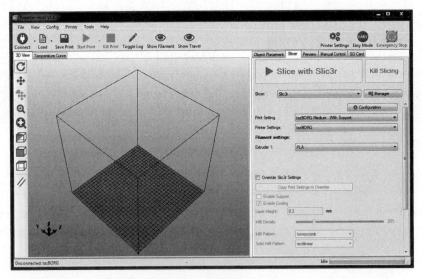

A screen capture of Slic3r, an open-source program for slicing models for 3D printing.

These horizontal cross-sections (called *layers*) can vary in thickness depending on the hardware of the 3D printer being used and the settings entered by the user. For consumer 3D printers, the layer thickness is generally somewhere between .10mm and .50mm, although it can be outside that range with the right hardware and settings.

Check to see if the manufacturer of a 3D printer has published recommended layer settings. Manufacturers usually do testing to determine an optimal layer thickness based on their hardware. These recommended settings will be a good starting point and will help you quickly achieve high-quality prints.

Layer thickness is one of the most relevant contributors in the quality of the printed part. The thinner each layer is, the higher the quality of print. The easiest way to visualize this is to picture it at its extremes. If you had a model with very thick layers, say 5mm (which is way outside of what actual 3D printers will print), your printed part would be extremely blocky. At the other end of the spectrum, a model printed with extremely small layers, like .01mm, would yield a very-high-quality result.

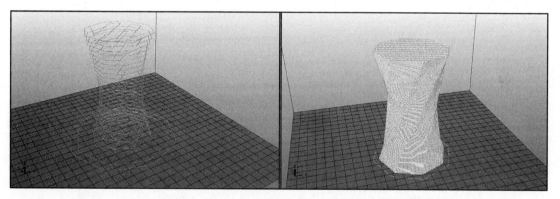

The difference between 5mm (left) and .01mm (right) layer heights.

But, of course, there are limitations that keep you from being able to print very small layers. The hardware itself, especially for consumer fused filament fabrication (FFF) 3D printers, has limits on how thin it can extrude filament. Even if the hardware was capable of extremely thin layers, it would take an impractical amount of time for the slicing software to generate the layers.

The most noticeable factor for the end user, however, ends up being print time. As the thickness of the layers decreases, the total number of layers increases. The more layers there are to print, the longer it will take the print to complete. For example, if a particular model takes two hours to print with layers set at .4mm, you would expect it to take four hours when the layer thickness is cut in half to .2mm. You doubled the number of layers, and so the time to print would double as well, right?

Unfortunately, that's not the case, and the print time would actually increase by a lot more than that. The reason for this is that the filament being extruded has a 3D profile, so as the height decreases (in the Z direction), its width (in the X/Y plane) also decreases. The result is that each layer itself takes longer to print, in addition to having more layers to print.

Extruding Filament

Filament extrusion is how the 3D printer actually creates each of those layers. In consumer FFF printers, the process generally works like this: a spool of filament is fed into an extruder (see Chapter 7), which has a motor and drive system to push the filament into the *hot end*. The hot end then very quickly melts the filament and pushes it through a small nozzle. The result is a thin thread of soft molten plastic being squirted out of the nozzle.

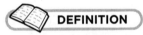 **DEFINITION**

> The **hot end** of a 3D printer is what heats and melts the filament that is fed by the extruder assembly. As filament is pushed into the hot end, a heating element heats the hot end. The temperature is hot enough to almost instantly melt the plastic into a very viscous fluid, which is then squeezed out of the nozzle and deposited on the print bed.

While the 3D printer is running, the tip of the nozzle is held very close to the print bed. As it's being ejected from the nozzle, the thread of plastic sticks to the print bed. The 3D printer continuously moves the tip of the nozzle around the print bed, trailing a thread of plastic behind, to form a solid layer of plastic that matches the layer generated by the slicing software. The printer then moves the nozzle up slightly (relative to the bed) and starts extruding a new layer of plastic on top of the previous layer. That process is repeated until all the layers have been printed and the part is complete.

The size of the molten plastic thread coming out of the nozzle, the distance between the nozzle and print bed, and the amount of filament being pushed into the hot end are how the 3D printer controls the thickness of the layer it's printing. The thread size is directly proportional to the size of the hole in the tip of the nozzle. So layer thickness can be adjusted by changing these parameters.

Chapter 3: Manufacturing with 3D Printers and CNC Mills

Molten filament being extruded to create the first layer of a part.

However, as I touched on in the previous section, the layer thickness is linked to the amount of time it takes to print the layer. As that thread of molten plastic gets thinner, it takes more time to create a solid layer. The result is a trend that holds true in most aspects of 3D printing: the quality of the printed object and the time it takes to print are locked together.

CNC Milling and How It Works

Milling machines are a standard tool in machine shops around the world. In their most basic form, they have a motor called a *spindle* that spins an end mill (which is very similar to a drill bit). The material to be milled is attached to a platform that can be moved in three (or more) dimensions, relative to the end mill. The machinist then manually moves the platform (or end mill) around using control mechanisms. This allows him to cut away at the material with the end mill to form the desired part.

 FASCINATING FACT

Before the advent of computer numerical control, there was just numerical control. Numerical control mills relied on a complex mechanical punch card system to program the mill. These programs took a great deal of time and expertise to create, but the repeatability of the programs made numerical control mills useful in manufacturing settings.

CNC mills simply remove manual operation from the equation. Instead of a machinist having to manipulate the controls, a computer controls the movement. The way the computer controls movement is actually pretty similar to the way 3D printers work: electric motors move the end mill in the X, Y, and Z axes relative to the print bed. The primary differences are the subtractive manufacturing aspect and the materials CNC mills can work with compared to 3D printers. The similarity between the three-axis movement in CNC mills and 3D printers is what makes it possible to convert a CNC mill into a 3D printer. It's also what allows manufacturers to make machines that can handle both tasks.

Subtracting with CAM Software

Because CNC mills create a part by subtracting material (as opposed to adding it like a 3D printer), the software and process they follow are completely different. While 3D printers add material in layers, CNC mills start with a block of material (or bar, cylinder, and so on depending on the stock used) and cut away material in "pockets." They don't cut away a layer at time, but rather cut out particular sections and features (often completely) before moving onto other features. Sometimes cast metal parts are even used as the initial stock and then are refined with CNC milling.

This is controlled with computer-aided manufacturing (CAM) software. Like slicing software, CAM software takes a 3D model (often in STL format) and processes it to create instructions for the CNC mill to follow. However, unlike slicing software for 3D printers, CAM software generally requires much more operator interaction. This is a profession unto itself, because setting up a CNC program requires experience and skill.

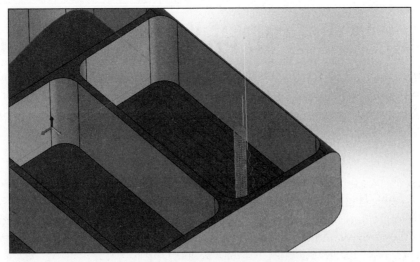

CAM software serves the same purpose as slicing software but follows very different rules.

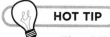

HOT TIP

The skill and experience of the CNC operator is important because CNC milling techniques and settings can vary tremendously based on the shape of the part being made, the material it's cut from, the type of end mill being used, and the desired surface finish and quality. For these reasons (among others), 3D printers are more suitable for home use, as the slicing software is much simpler to use than CNC CAM software.

Milling the Material

While 3D printers are mostly limited to plastic (except in experimental printers and very expensive printers), a common CNC mill is capable of creating parts from a variety of materials. Steel, aluminum, brass, titanium, wood, and most plastic types are all possible to machine on a suitable CNC mill. Whatever the material, CNC mills start with a block of it, and whatever is milled is either thrown away or recycled.

That versatility makes CNC milling useful for more than just rapid prototyping. CNC mills are very often used to actually manufacture parts, especially aluminum and steel parts. The precision of CNC milling also allows it to achieve very tight tolerances that are necessary for production parts.

However, because milling is a subtractive process, the material that's cut away is generally wasted. 3D printers usually only use the material that's actually needed to create the part (although sometimes more is used for support structures). Considering the high cost of materials, that can be a significant factor in the price of the part.

Pros and Cons of 3D Printers vs. CNC Mills

3D printers and CNC mills are both versatile and useful, and each excels at different tasks. Each has jobs for which they're better suited, and each has jobs for which they would be a poor choice. So what are the pros and cons of each? When should you use one versus the other?

Cost

The most obvious difference, for anyone purchasing a 3D printer or CNC mill, is going to be cost. Capable consumer 3D printers can be purchased for under $1,000, while CNC mills capable of machining metals like aluminum and steel are rarely less than $5,000 (and are usually much more). However, less powerful CNC mills capable of milling soft materials like wood are significantly less expensive (usually less than $2,000).

The cost difference is mostly due to the expensive high-power electric motor needed for the CNC spindle, the cooling systems needed when milling metal, and the fact that CNC mills require a very solid and rigid frame. 3D printers, in comparison, require much less expensive parts to build.

Aside from the price of the machine itself, there are other costs to consider. CNC mills require end mills, which need to be replaced after they become worn. Most CNC operators prefer to have a range of end mills as well, in different sizes and shapes for different jobs. 3D printers, on the other hand, generally only require that your purchase the material itself.

If you wanted to convert a CNC mill to a 3D printer, it is relatively inexpensive, generally only a few hundred dollars.

HOT TIP

While 3D printers don't require expensive consumable accessories like end mills, you should consider the cost of filament. Most people end up purchasing many rolls of filament in a variety of colors and materials. This expense can be significant, especially for exotic filament materials.

Part Geometry

One of the biggest advantages that 3D printers have to offer is their ability to create complex geometry. There are parts that can be 3D printed that simply aren't possible with traditional manufacturing methods. The way that 3D printers create parts in layers means that internal geometry can be produced just as easily as external geometry, a prospect which simply isn't possible with any other manufacturing method.

An easy-to-understand example of this is a simple cube with a hollow sphere in the center. No other manufacturing method is capable of producing such a part, but it's a trivial task for a 3D printer. Practically speaking, it's not any more difficult for a 3D printer to create a part like that than it would be to create a solid cube.

This capability unleashes a whole world of possibilities that were previously considered unfeasible or simply impossible. You could even print two (or more) parts that are already assembled, allowing you to create assemblies which wouldn't be possible with any other manufacturing tool.

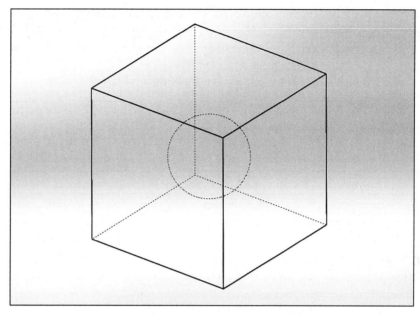

A part that could only be created by a 3D printer.

Even more traditional parts are easier to create with a 3D printer. Parts with features on all sides, for example, can't be machined on a standard three-axis CNC mill without additional work. The part would have to be turned over to machine the other side, or a fourth axis would have to be added to the mill for rotating the part (which is a fairly expensive addition).

However, 3D printers do have limitations on what kind of geometry they can handle (especially consumer FFF printers). These limitations are usually related to overhangs in the part, as there always has to be something to print on top of. But that can be overcome by printing support material which is later removed.

FASCINATING FACT

3D printers with two or more extruders can use one of the extruders solely for printing support material. This support material is often a special kind of filament produced specifically for this use. Once the print is finished, it can be soaked in a chemical bath, which dissolves the support material but leaves the rest of the part untouched. Such a setup would be capable of producing almost any geometry imaginable.

Material Matters

As I explained in the previous sections, the material CNC mills are capable of machining is a lot more varied than what 3D printers can print. 3D printers in the consumer market are limited to different kinds of plastics, while CNC mills can machine most common materials (with the right setup).

But it's actually not that simple in practice. The cost of material is certainly a concern. Machinable blocks of material suitable for CNC milling tend to be fairly expensive and have to be purchased or cut in the necessary sizes. If the part you're making isn't very close in size to the block of material, the chances are good that a lot of material will go to waste.

Milling different materials also requires that the CNC mill be set up for that material. The end mill, cooling system, and even the frame of the CNC mill itself needs to be correct for that material. This is why mills are less expensive when they only need to be capable of cutting soft materials, and are more expensive when they need to cut hard materials like steel. Each material also requires different milling settings, and it takes experience and knowledge to properly mill a variety of materials.

In contrast, 3D printers generally only require that you adjust a few settings (namely temperature) when switching between materials. Simply put, 3D printing is much easier, and the learning curve is shorter than CNC milling. CNC mills certainly have the advantage of material versatility, but that versatility is difficult to achieve.

Surface Finish

The final quality of a part is largely a consequence of the surface finish on the part. The surface finish, in this context, is mostly about how smooth the completed part is. In a manufacturing setting, surface finish has a slightly different connotation. For production parts, a smooth surface isn't always what is desired. Sometimes it's preferable to have a rough finish, brushed finish, or matte finish. These are all different types of surface finishes, and there are many others. No particular type is necessarily better; it's just a matter of what's needed for the part.

However, in most cases, the manufacturer must first start with a smooth finish and then add the desired surface finish. If we use injection molded plastic parts as an example, this process is part of making the mold. If a textured matte finish is desired on the part, that finish has to be on the surface of the mold. In this case, the mold is milled as smoothly as possible, and then the finish is added afterward (either mechanically with sanding or with chemical etching).

In almost every case, the mold or part is ideally as smooth as possible to begin with, and then the surface finish is added. The reason it's done that way is to avoid imparting unintended artifacts onto the surface. And so, you'd want to be able to achieve as smooth of a surface finish as possible when initially making a part.

Getting a very smooth finish is usually a simple matter with a CNC mill. It does take skill on the CNC operator's part, and usually a smoother finish takes longer to machine, but it's certainly possible to get a very smooth finish straight off the mill. Unfortunately, this isn't the case with 3D printers, and is probably one of the primary reasons that 3D printing isn't more common in production manufacturing.

As I've explained previously, the nature of 3D printing makes ridges on the surface almost unavoidable. The way 3D printers create parts by adding layers virtually ensures that small differences between those layers will be visible. Very-high-end (non-FFF) 3D printers used in professional settings can achieve smooth surface finishes, but those kinds of printers are far out of reach for consumers. Their high cost also makes them impractical for any kind of manufacturing, even for the companies that can afford them.

Even at high-quality settings, a 3D-printed part has a rough surface.

So is it ever possible to achieve a smooth surface finish on a consumer 3D printer? Luckily, the answer is yes! Smooth finishes on 3D-printed parts can be attained with postprint finishing. There are many techniques to do this: everything from simply sanding the part to more complicated chemical methods. But those postprint finishing methods are still an additional step that CNC mills don't require.

At the end of the day, if you're looking for professional-looking parts with a smooth surface finish, CNC mills are the better choice. Without postprint finishing, 3D-printed parts will always look like they were 3D printed. You can get a nice and professional finish on a 3D-printed part; it will just take more work.

Using the Right Tool for the Job

With all of these differences, how do you know if you should use a 3D printer or a CNC mill? As always, it's important to choose the right tool for the job. Just like you wouldn't use a wood saw to cut a metal pipe, you don't want to use a 3D printer when a CNC mill would do the job better (and vice versa). In this case, choosing the right tool is about first determining what the job will be.

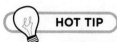

> There are practical differences which might affect your decision as well. For example, CNC mills are very loud and can be quite messy. 3D printers, however, are usually fairly quiet and don't create much mess. If you live in an apartment, a CNC mill would probably be pretty disturbing to your neighbors, while a 3D printer shouldn't even be noticeable to them.

If you intend to make production parts or molds for injection-molded parts, a CNC mill is generally the clear choice. They're still capable of creating prototype parts in a short amount of time but can also produce professional parts with a high-quality surface finish. However, their noise, mess, and expense are certainly a concern.

For most home users and hobbyists, a 3D printer will be easier to use and more useful in general. You can create a wide range of parts with little concern about geometry limitations and complicated settings. The cost of material is lower, you waste less of it, and the noise and mess are minimal.

In the end, it's easy to see why a machine capable of both tasks would be so useful. The frame, movement mechanics, and controls are very similar between 3D printers and CNC mills. Switching the CNC spindle for an extruder allows you to do both CNC milling and 3D printing with the same basic machine. This is certainly more expensive than a dedicated 3D printer or CNC mill, but much less expensive than purchasing one of each.

The Least You Need to Know

- 3D printers create parts by building up layers of material, a process called *additive manufacturing*.
- CNC mills use subtractive manufacturing, which means they create parts by cutting away material.
- Generally speaking, 3D printers are better for the home user, while CNC mills may be a better choice for businesses, depending on the expected use.

CHAPTER 4

Types of 3D Printers

In This Chapter

- Stereolithography and its subset, digital light processing
- Creating colorful models with powder bed printers
- Printing strong parts with MultiJet printers
- Melting particles with selective laser sintering
- The reason for fused filament fabrication's popularity

So far in this book, I've talked about 3D printing in generic terms that apply to most 3D printers. But in practice, not all 3D printers work in the same way. To move forward with this book, I need to pick a particular 3D printing process that I can dig deeper into.

There are many kinds of 3D printing technologies on the market today, and each of them operates differently. Most of these processes are used specifically for a certain application or a particular material. I don't have the space to cover every process here, so instead I go over the most common types of 3D printing in use today, plus talk about why fused filament fabrication (FFF) printing is the go-to type for many hobbyists.

Stereolithography

If you recall from Chapter 2, this is the original type of 3D printing process developed by Charles W. Hull and his company 3D Systems in the '80s. Stereolithography (SLA) works by shining a UV laser onto a vat of UV-curable photopolymer resin. The laser is focused on the resin to produce the 2D cross-section of each layer on a build platform, and then the platform is slightly lowered in order to form the next layer.

SLA, aside from having the proven history of being the first 3D printing process, has some key advantages which make it a popular choice even today. An SLA printer can print parts with relatively smooth surface finishes compared to other types of 3D printing, with almost completely unnoticeable differences between layers. This quality is possible because of the high precision of the UV laser, which also allows it to create parts within very tight tolerances.

The quality and precision of SLA printers makes them ideal in professional research and development situations, where prototype parts need to very closely resemble their mass-produced counterparts. The surface finish also has a high-enough quality that SLA printed parts can be used as masters for creating molds.

However, the quality of an SLA printer comes at a high price. High-end consumer model SLA printers have started entering the market recently, but they've mostly been used only by businesses for creating prototypes. The cheapest SLA printers start in the thousands of dollars and can be more than a hundred thousand dollars for professional models.

Aside from the cost of the printer itself, expensive resin is also needed as the material to create parts. This resin can cost anywhere from $80 to $200 per liter, and is much more expensive than the plastic filament used in most consumer 3D printers. The way SLA printers create parts in a vat also means that, for conventional designs at least, hollow parts will be filled with resin. This can add to the cost of the material and to the weight of the part.

In some cases, the printed part requires additional curing as well. This is dependent on the printer and resin being used, but is still a factor. Curing the part generally requires a separate device and additional time.

Digital Light Processing

Digital light processing (DLP) 3D printing is very similar to SLA printing; in fact, it's considered a subset of SLA printing. DLP printing works almost exactly like SLA printing, with a vat of photopolymer resin that is hardened layer by layer.

The primary difference between DLP printing and traditional SLA printing is the method used to shine light onto the resin. If you're familiar with video projector systems, you may have already heard of DLP in the context of movie projects. The technology is used to project light (in the form of images) onto a screen.

This same technology can also be used to cure the layers of resin in a DLP 3D printer. Instead of projecting a video, a DLP printer projects light in the shape of each cross-section on the resin. This allows a DLP 3D printer to cure an entire layer at once, a big advantage over the single-point laser used in conventional SLA printers.

FASCINATING FACT

DLP 3D printers are capable of curing an entire layer of resin in just a few seconds. This speed allows them to print very quickly compared to other types of printers.

By curing an entire layer at once, DLP printers dramatically reduce the time it takes to print a part. Additionally, the size, shape, and complexity of each layer have no effect on the time it takes a print. Aside from the obvious advantage of being able to print large and complex parts, this also means that many duplicate parts can all be printed at once without adding any print time compared to a single part.

However, DLP 3D printers tend to have the same downside as standard SLA printers: cost. While the price of resin is usually about the same as SLA resin, the printers themselves are more expensive. DLP projectors are still complex and expensive pieces of technology, and adding them to a 3D printer is not a cheap proposition.

Powder Bed Printing

A handful of specific printing processes could all be considered powder bed printing. All of these types of printers have one thing in common: a powder material is spread on a bed, and a liquid binder is used to solidify it in layers. The differences come from the following:

- The type of powder material used
- The type of binder used
- The method used to inject the binder into the powder

The types of powder and binder used can vary based on need, and have evolved over the years since power bed printing was invented. It can be anything from simple plaster powder that is solidified by injecting water (like the original binder jetting process developed at MIT and sold by Z Corp.) to advanced resin or epoxy combinations that replicate common engineering materials.

One of the most common and popular of these powder bed printers on the market today is the inkjet type. They use a print head similar to what is used in common household inkjet printers, but instead of printing ink, they print a binder onto the powder.

There are two primary advantages that powder bed inkjet printers have over other types of printers: support material isn't necessary, and they have the capability of printing multicolor parts. They don't need additional support material because the powder itself can support overhangs while the part is being printed.

The leftover powder can be reused, so not much material is wasted. However, cleaning the powder off the part isn't a small job. It can be a time-consuming task, sometimes made even more difficult by the fragility of the part produced by some printers (and depending on the material). Removing the powder without damaging the part in those cases takes finesse and care, but the other major benefits can make that inconvenience worthwhile.

By using multiple inkjets in the print head, each with a different-color binder, parts can be printed in multiple colors. Colors can even be combined, similarly to inkjet printing on paper, to create a wide spectrum of colors.

The ability to handle overhangs with ease and to print colorful parts has made powder bed inkjet printers very popular for creating architectural models. Usage for architectural models is also a result of the relative ease of building large powder bed printers, which aren't limited to small sizes like other technologies. This allows architecture firms to print large full-color models of their building designs for presentation purposes.

Most powder bed printers have a major downside though, which is that prints tend to be brittle and weak (though this problem has largely been solved with newer materials and processes). This often necessitates the use of additional postprint treatments just to make the part suitable for handling. The powder bed also introduces the same problem that is common with printers that use vats of resin: you can't print a hollow part without a way of removing the unused material inside before it's completely sealed.

MultiJet Printing

MultiJet Printing (MJP), also called *PolyJet Printing*, is another 3D printer technology that uses an inkjet-style print head and can print in multiple colors. But unlike powder bed printers, MJP printers don't use powder beds; instead, they print UV-curable resin directly onto the print bed. MJP printers add a UV light to the print head, which instantly cures the photopolymer resin as it's deposited.

HOT TIP

MultiJet Printing and PolyJet Printing are examples of different companies using similar technologies under different names. MultiJet Printing is used by 3D Systems, while Stratasys uses PolyJet Printing. Both work in approximately the same way but use different names for trademark, patent, or marketing reasons. When looking at 3D printers, keep in mind that the same type of printer can go by multiple names.

Because they use resin instead of powder, MJP and PolyJet printers can produce strong and usable parts. This gives them a distinct advantage over powder bed printers, which produce relatively weak parts.

MJP printers handle overhangs by printing a gel support material where necessary. The support material can be easily washed away without damaging the part, making the support removal process easier and faster than more traditional approaches.

The disadvantages of MJP and PolyJet printers are the cost of the printer itself and the cost of the resin. The resin cost is on par with the resin used in other types of printers. However, to print in multiple colors or materials, different types of resin are required. In the long term, this shouldn't add any significant cost, because the total amount of resin being used should be roughly the same. But in the short term, it can be a large upfront cost to get outfitted with a range of resin types.

The printers themselves are priced similarly to other professional 3D printers, which of course are still too expensive for most consumers. Being a new technology, MJP printers haven't yet matured enough to reach more affordable prices that would allow them to be purchased for home use.

Selective Laser Sintering

Selective laser sintering (SLS), direct metal laser sintering (DMLS), and selective laser melting (SLM) all work in basically the same way. They use high-power lasers to actually "sinter" or melt powdered particles together into a solid mass.

Like power bed inkjet printers, they use a container that is filled with powder layer after layer. After a new layer of powder is added, the laser heats the part cross-section and then moves onto the next layer. As usual, the process is repeated until the part is finished.

However, unlike other powder bed printing methods, SLS printing doesn't require any kind of binder. That means any material that can be powdered and melted or sintered with a laser can be used, such as various metals, ceramics, plastics, and even *green sand*.

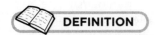

DEFINITION

Green sand is a specially formulated material used for sand casting. Sand casting is used to produce metal parts by pouring molten metal into a mold made of sand. The mold is normally produced by forming it around a positive master part; however, 3D printing the mold removes the need to first create a master part.

These are the only commonly used 3D printers capable of printing metal parts, which gives them an obvious advantage. In some cases, the printed parts can even be as strong as an identical part manufactured by traditional means. This makes SLS-created parts particularly unique and especially useful.

Unfortunately, SLS, DMLS, and SLM 3D printers are all very expensive at this point, to the point that they're mostly only used in a handful of industries that require the specific capabilities they offer. In most cases, a capable multiaxis CNC mill is more economical than an SLS printer, so companies that purchase them don't usually do so without a strong necessity.

Fused Filament Fabrication

All of the technologies I've talked about so far in this chapter have been too expensive or too specialized for consumer use. So what process are you likely to use at home? That's where fused filament fabrication (FFF) comes in. This process takes a string of thermoplastic (called *filament*), melts it, and deposits it on to the print bed.

FFF printers are highly economical, making them popular for consumer use. The affordability of the printers themselves is due primarily to the low cost of the individual parts which make up the printer and the fact that they don't need to be precisely built to function well. Most of the other printer types I've discussed are difficult to build and have to be finely tuned to produce good results.

On the other hand, FFF printers can be built in a garage with common hand tools and still print well. Mechanically speaking, they're also relatively simple and easy to understand. No complex laser or optical systems are needed, because they operate on mechanical principles.

If you've ever used a hot glue gun, you can probably understand how an FFF 3D printer works. The filament is comparable to the glue stick, the hot end is similar to the heated nozzle of the glue gun, and the extruder works like the trigger system that pushes the glue stick into the hot nozzle (although extruders work with continuous rotation).

The FFF 3D printer process, then, is like using a hot glue gun to draw a square, and then another square on top of that one, and then another square on that one, and so on. The 3D printer is doing it much more precisely than you could by hand, of course, and it's doing it with plastic, but the general idea is the same.

While the affordability of FFF 3D printers makes them ideal for consumer use, the technology is used for professional 3D printers as well. Professional versions work in exactly the same way as consumer models, just with greater precision and sometimes with additional features added.

The cost advantage doesn't stop at the printer itself either. The filament used in FFF printers is the most inexpensive material available for any 3D printer out there. This is partially due to the low-tech nature of the filament, but it's also a result of the competition in the consumer market driving prices down. The popularity of FFF printers has also ensured rapid development of both the printers themselves and the filament used with them.

While FFF printers are popular mostly because they're affordable and easy to understand, unfortunately, they probably yield the lowest-quality prints of any of the 3D printer types discussed in this chapter. The filament extrusion method of layering plastic is inherently imprecise. And, while a great deal of effort is being put into improving the quality, FFF printers still lag behind the others.

HOT TIP

While the quality is lower than other printers, FFF printers can print a range of thermoplastics. Many of those thermoplastics are popular for engineering work because of their superior mechanical properties. Parts printed with materials like ABS and nylon are very strong, and are usable as soon as they're finished printing.

Another downside is the time it takes an FFF printer to create a part. Because of the weight being moved around, FFF printers can only accelerate so fast. This makes small features slower to print. Because the plastic is melted as it's being deposited, it's also necessary for it to cool before printing another layer on top of it. Most of the time, it's already cool by the time the next layer is printed. But if the layers are very small, an FFF 3D printer will either need to pause between layers or risk deforming the part due to the heat.

That heat also provides the underlying cause of the single biggest disadvantage of FFF printers: warping. Because thermoplastics expand and contract as they're heated and cooled, the part actually changes shape slightly as it's being printed. If some layers cool before the others, this will result in the part warping or even cracking.

An FFF-printed part showing warping and cracking.

How extreme this warping is depends on the particular thermoplastic being used. Acrylonitrile butadiene styrene (ABS), for example, is especially prone to warping and cracking. Other materials minimize warping, but none of them seem to be able to avoid it entirely.

Printer manufacturers, in an effort to avoid warping and cracking issues, have introduced a number of ways to try and control it—heated beds, enclosures, and heated print chambers. These methods are mostly related to managing the temperature of the part to try and keep it consistent throughout the print. These work fairly well and almost completely eliminate the problem in some materials. Right now, however, it still can't be avoided entirely.

FASCINATING FACT

Some of the methods of reducing warping, such as heated build chambers, are patented technology. This means that small 3D printer manufacturers can't use those methods without making a deal with the patent owner. This is part of the reason that some of the features seen on professional 3D printers aren't available on consumer printers.

Though warping problems and relatively low print quality are certainly considerable disadvantages, there is no denying the low cost of FFF 3D printers. They are by far the most popular type of 3D printer on the consumer market, and virtually all printers in use by hobbyist and home users today are FDM/FFF 3D printers. Therefore, it's extremely likely that this will be the type of printer you'll be purchasing.

For that reason, throughout the rest of the book, I'll be talking about 3D printing in the context of FFF 3D printing. Many of the principles apply to other types of 3D printing as well, but the FFF process is what I'll be focusing on.

The Least You Need to Know

- There are many types of 3D printers on the market, each with their own advantages and disadvantages.
- While SLA, DLP, powder bed, MJP, and SLS printers print high-quality parts, their cost keeps many hobbyists from buying them.
- FFF 3D printers are the most popular for consumer use because of their affordability.

PART 2

All About the Hardware

3D printers are complicated machines, which is why I'm devoting this part to explaining how all of that hardware actually works. Not only does understanding the hardware help you understand how 3D printing works, but it also gives you the knowledge necessary to troubleshoot any problems you may run into. You also learn how best to prepare your 3D printer for successful prints.

CHAPTER 5

The Frame

The frame of a 3D printer is what everything else is built on, making it a very important factor in the quality of your prints. Printers are built in a variety of different ways, and all the different layouts and construction materials make a difference. To understand why, it's best to start by learning how 3D printer frames are generally constructed.

In this chapter, I go over the Cartesian layout of 3D printers, construction techniques for them, and the importance of size when it comes to your 3D printer.

In This Chapter

- Cartesian-layout 3D printers
- The importance of frame rigidity
- Choosing the right-size printer for you

Cartesian Layouts

When it comes to 3D printers, *layout* refers to the way the printer is designed to move in three-dimensional space. That movement can be achieved in different ways, but the most common of those is the *Cartesian* style. These printers are named after the Cartesian coordinate system, which is the most basic way to define a point in 3D space. This coordinate system is what you probably remember using in your math classes in school: a point is defined by its X coordinate and Y coordinate on a 2D plane. With 3D Cartesian coordinate systems, you're simply adding another axis (Z) to define the vertical position of the point.

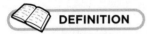

DEFINITION

Cartesian is an adjective used to describe things related to René Descartes, who was a French mathematician and philosopher. His many contributions to mathematics were the reason for the Cartesian coordinate system being named for him (though he wasn't solely responsible for its development).

Cartesian 3D printers use the same basic principle to move the tip of the hot end to a specific point. They have mechanisms to move in the X axis, Y axis, and Z axis, allowing them to position the hot end anywhere in 3D space. For example, if the hot end needs to move from the point [10, 10, 10] (X, Y, Z) to [20, 5, 10], it's a simple matter of moving the X axis 10 units in the positive direction and the Y axis 5 units in the negative direction.

3D printers with Cartesian layouts are by far the most common, both in the consumer market and the professional market, most likely due to the simplicity of the math involved in controlling their movement. Other types of layouts, like Delta printers, require the use of trigonometry to adjust for a simple move on a single axis. Cartesian printers also seem to be easier for people to understand because they mirror the Cartesian coordinate systems used in CAD programs (see Chapter 15).

Ways Cartesian 3D Printers Are Constructed

Despite the simplicity of the way Cartesian 3D printers handle positioning, their actual construction can be quite varied. The X, Y, and Z movement is only how the hot end moves relative to the print bed. The parts that are actually physically moving can be completely different between printers. One 3D printer model might have the print bed move in the X and Y directions, and the hot end move in the Z direction. Another printer might have the print bed move in the Y direction, and the hot end move in the X and Z directions.

Chapter 5: The Frame   51

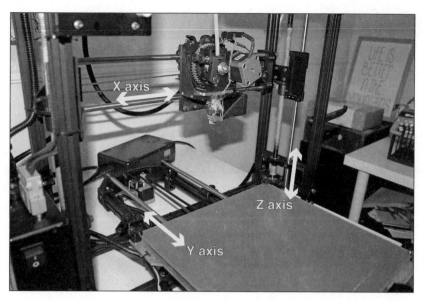

This 3D printer has a bed that moves in the Y axis and an extruder carriage that moves in the X and Z axes.

The most common setup right now seems to be to have the print bed moving in one direction (either X or Y) and the hot end moving in Z and the other direction (either X or Y). But this is by no means a rule, and it's not even necessarily the best method. That's mostly because what's best depends on your priorities and what you're trying to achieve.

Cartesian Layout Considerations

So why do 3D printer manufacturers use different Cartesian layouts? Should it even matter to you? The short answer to the first question is that one 3D printer manufacturer might be trying to achieve something different than another manufacturer. The particular layout they choose to use can affect the cost of the printer, the quality of the prints, the speed at which it can print, the overall size of the printer, and the difficulty of building it.

Cost is usually the most obvious of the factors, and it's certainly noticeable to both the manufacturer and the customer. The cost varies between different setups because of the materials needed and how powerful the motors need to be. If all of your movement is done by the hot end while the print bed stays stationary, the Z axis motors will need to be very powerful in order to lift the weight of all the components of the X and Y axes. And of course, if a particular layout uses more material in its construction, it will cost more.

The quality of the prints and the speed at which the printer can print are also inherently interrelated. Generally speaking, print speed is limited by print quality. Most printers are capable of physically moving much faster than they actually print. It's a matter of how fast a printer can move while maintaining acceptable print quality. These two things are largely determined by the amount of mass being moved and how well the frame of the printer can handle the stress of that mass.

Print quality is affected by moving mass because of simple inertia. If you remember your physics lessons, you'll recall that inertia increases proportionally with mass. Inertia is the resistance of a mass to change in its current state of motion. The reason this affects print quality is pretty straightforward: if you're printing in one direction and need to change direction (for example, when you reach a corner), inertia will resist that change. And because inertia is related to mass, that change in direction will be more difficult as more mass is added. So the printer will have a tendency to overshoot the corner, resulting in poor print quality on that corner.

This is why speed is a significant factor as well. If the moving mass is high on that axis, the printer may not be able to turn that corner quickly at high speed. But at lower speeds, the effect will be reduced, resulting in better print quality. The lesson here is that you can print at faster speeds as moving mass is lowered while maintaining the same print quality.

FASCINATING FACT

Backlash (an undesirable delay in movement during direction changes) produces somewhat similar effects to those caused by poor rigidity, but is a separate phenomenon with a different cause. Backlash is caused by looseness in the interfacing parts of the linear movement systems used on 3D printers. When that looseness is present, there is a small delay before the system is engaged when the direction of an axis's movement is changed.

But what does that have to do with the particular layout used for a given 3D printer? It means that printer manufacturers try to reduce the moving mass on the X and Y axes, which is why that movement is often divided between the hot end and the print bed. That way, the one axis only has the mass and inertia of the extruder to deal with, while the other axis only has to handle the mass and inertia of the print bed. This allows you print at higher speeds while maintaining an acceptable print quality.

The size of the printer is also directly affected by the layout chosen, especially when you consider the total area needed by the printer when the print bed moves. If the print bed were to move in both the X and Y directions, the printer would need four times the area of a completely stationary bed. This is simply because each axis would need to be twice as long as the bed in order for the nozzle to reach every point on the bed. So if the goal is to produce a very compact

3D printer that takes up very little desk space, a stationary bed with all movement done by the hot end would be ideal. Of course, that would lead to a lot of moving mass, and potential print speeds would be lower.

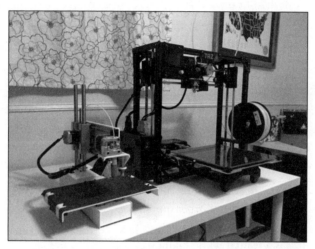

The size difference between a Printrbot Simple (left) and a LulzBot TAZ 4 (right). Note, however, that the Printrbot has a 6×6-inch bed, while the LulzBot has a 12×12-inch bed.

How all of this affects the difficulty of actually building the printer should be readily apparent. The more complex the design of a 3D printer, the longer it will it take the manufacturer to assemble it (or you, if you buy a kit). Between this and all of the factors involved, it's obvious that 3D printer manufacturers have a lot to consider when designing a printer.

But should you be concerned with what layout a particular model uses? It does matter, for the reasons I've gone over in this section. But it's difficult to determine real-world results based on the layout alone. That's because things like the inertia and momentum can be counteracted, and one of the best ways to do that is with a high-quality frame.

The Importance of Frame Construction

How the 3D printer is actually constructed is one of the most important factors when it comes to print quality and reliability. Every other part of the printer could be perfect, but if the frame is poorly constructed, the results will be very poor.

A 3D printer frame doesn't just need to hold the other parts together; it also has to keep them stable and aligned. It has to hold up to the forces of momentum and inertia caused by the mass being moved around the printer's axes, as well as keep the axes properly aligned at all times and maintain the calibration of the printer under the stress of constant movement and vibration.

Rigidity and How It Affects Quality and Reliability

The single most important characteristic of a 3D printer frame is rigidity. Flexibility in the frame is the biggest enemy of print quality. If there is any flex in the frame, the momentum of the moving parts will result in poor print quality, unless you print at slow speeds.

Reliability is also affected by the rigidity of the frame. 3D printers require calibration in order to produce high-quality prints. Calibration involves, among other things, setting a Z height in a "Goldilocks zone." If the Z height is too high, the first layer won't adhere properly to the print bed; if it's too low, you won't be able to extrude a clean and solid line.

The acceptable range for Z height is very small; usually, it needs to be within $\frac{1}{20}$ of a millimeter for good results. A Z height outside of that range will result in failed or poor-quality prints. To keep from having to frequently calibrate the Z height, a 3D printer should be capable of maintaining your calibration for a long time. A flexible or loose frame will cause your Z stop, Z axis components, or hot end to move slightly over time. That slight movement means that you will have to constantly recalibrate your printer to continue to get good results.

What Makes a Good Frame

So you now know why it's important for the frame to be rigid, but what makes a high-quality and solid frame? Rigid frames have a few characteristics in common:

The design of the frame structure: 3D printer frame structure design is a fairly complex subject. Because of the complexity involved in designing a frame structure optimized for rigidity, a lot of 3D printer manufacturers take the overkill approach by using heavy-duty materials for the frame. The best way to be sure you have the most rigid frame possible is to make it fully boxed, with each frame piece connected at both ends to another part of the frame. Other more unconventional designs can still yield good results, but difficulty of engineering them makes them less common.

 HOT TIP

Your intuition might be your best tool when it comes to determining the quality of the frame. Ideally, you want to actually get your hands on the printer and see how it feels. If you can flex it with your hands, it's probably not a good choice. But even just looking at it should give you a good idea: if it looks nice and sturdy, it probably is.

The connections between frame components: Even the most ingenious frame design is going to fail if the frame is held together with duct tape. Frame pieces should be connected together with a strong material that resists flex, while the connections themselves should be designed to resist movement in all directions. A simple flat sheet metal 90-degree connector will do a great job of resisting flex and movement parallel to its plane but will do a poor job when a perpendicular force is applied to that plane. To resist forces in all directions, connectors should be 3D instead of flat. Or, if flat connectors are used, two should be used together in a perpendicular orientation.

The frame material: In essence: strong materials are good. Metal is better than wood. Steel is stronger (but heavier) than aluminum. Plastic could be okay if it's a high-quality plastic and the frame is well designed.

The best possible performance would be achieved with a big, heavy, cast-iron frame. These are the kinds of frames used in heavy-duty machine tools, because they're incredibly rigid and strong. However, the weight and cost make them pretty impractical for use in 3D printers (especially for consumer desktop printers).

A safe and common frame material is standard t-slot aluminum extrusion. This is a high-quality structural material used in a variety of industries. It's popular in 3D printer construction because it's fairly inexpensive, easy to find in all kinds of sizes, easy to work with and to connect parts to, and pretty strong for its weight.

T-slot aluminum extrusion is a very versatile construction material used in a wide variety of applications, including 3D printer frames like this one.

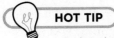

HOT TIP T-slot aluminum extrusion comes in a variety of sizes and shapes. 3D printers commonly use the very popular 20×20mm size, which is adequate for the application. But bigger would be even better to increase the strength of each piece of extrusion (assuming the overall design is the same).

Wood is also used pretty frequently, mostly because it's cheap and easy to laser cut and work with. However, wood probably isn't the best material to use in 3D printing. It's hard to achieve and maintain dimensional accuracy in wood parts, especially because they can expand or contract in the presence of moisture. Wood also tends to be at least somewhat flexible, which, as you know by now, is a bad thing.

Plastic and sheet metal frames can both be acceptable as long as they're well designed. They should be fully boxed and preferably reinforced. Plastic frames should be made from a sturdy and rigid plastic, making acrylic a popular choice.

Size Matters

Now that I have all of that dense engineering stuff out of the way, I can move on to a more straightforward topic: the size of the printer itself.

There is no denying the benefit of a large 3D printer. The bigger the printer, the bigger the parts you can print. Having the ability to print large objects is certainly useful. However, the size of a 3D printer influences more than just the size of the objects you can print.

The larger a printer gets, the more it's going to cost. This is partially because of the obvious material increase, but that's not the only reason. Bigger printers mean more mass, which means more powerful motors are needed for movement. That additional mass also means the frame needs to be stronger to resist flex. The longer smooth rods (see Chapter 6 for more on smooth rods) also need to be thicker so they don't sag.

Those powerful motors and a larger heated bed are also going to need more power. Not only does that mean you need a bigger power supply to feed them, but it also means you may need special control electronics that can handle the load.

Put that all together, and there are a lot of costs associated with increasing the size of a 3D printer. Subsequently, size usually ends up being the single biggest factor in the price of consumer FFF 3D printers.

Of course, price might not be the only downside to large 3D printers. Depending on how much space you have available, you just might not have enough room for a big 3D printer. A small printer that sits comfortably on the corner of your desk might be more suitable.

In order to figure out what size printer you need, you should ask yourself what size objects you're likely to print. Big parts can take a very long time to print, use a lot of material, and increase your chances of some kind of print failure during the print—there is nothing worse than a 24-hour print getting ruined 1 hour before completion. But the usefulness of a large print area is hard to deny. The ability to print large parts when needed is very handy, and it will be equally capable of printing small parts the rest of the time.

HOT TIP

No matter what size printer you're looking at, make sure the frame is nice and sturdy. Pay attention to the layout and look for any potential flaws, like the print bed or hot end moving on both axes. Look at the printer as whole, not just the individual components, to determine its quality.

The Least You Need to Know

- Cartesian 3D printers have mechanisms to move in the X axis, Y axis, and Z axis, allowing them to position the hot end anywhere in 3D space.
- Rigidity is very important in 3D printer design. Frames with flex will have poor print quality or will only be able to print slowly.
- The larger the printer, the bigger the parts you can print. However, larger printers tend to cost more.

CHAPTER 6

Movement Components

The frame of a 3D printer is the foundation for every other part, which makes it very important. But even the best frame in the world is useless without some kind of mechanical system for movement. The goal for all 3D printers is to provide linear movement on each axis. The specific way this is accomplished can vary based on the particular printer, but most use the same fundamental parts and systems.

In this chapter, I go over the pieces that accomplish the movement needed with a 3D printer.

In This Chapter

- How linear motion works in 3D printers
- Using stepper motors for precise movement
- How threaded rods make 3D printers more affordable
- Basic connection components

Components for Smooth Linear Motion

With any kind of linear motion system, it's important that the motion be constrained to a single axis in one dimension. Any kind of movement outside of that one dimension is almost always considered to be undesirable, because it introduces unintended movement in the other axes.

Furthermore, the motion on that one axis should be as smooth and frictionless as possible. The less friction there is in the system, the less force is required for movement on that axis. Friction in the system increases the power required for motion on that axis and can create inconsistencies in the movement.

For smooth linear motion, regardless of the particular machine, the most common setup is a smooth rod or rail with bearings on the moving part. This works in exactly the way that you would expect: the bearings reduce friction in the movement, and the rod or rail gives the bearing a smooth and straight surface to ride on. The concept is simple, but the individual components are still important.

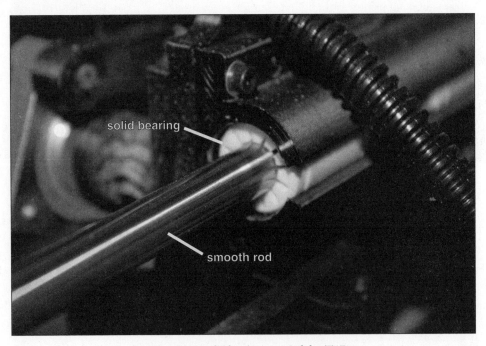

Smooth rods and solid bearings on a Lulzbot TAZ 4.

Rails and Smooth Rods

Smooth rods are exactly what they sound like: a straight cylindrical metal rod that is polished smooth. Rails are similar, except they don't need to be cylindrical and can be flat instead. Rails are often used when they need to provide some level of structural support, while rods generally aren't used as part of the structure. However, rods have the advantage of constraining movement in two directions with a single bearing, while rails often need two or three bearings to do the same.

FASCINATING FACT

In some 3D printer designs, the aluminum extrusion that makes up the frame is also used as the rail for moving parts to ride on. One or more bearings make contact with the surface of the extrusion, allowing parts like the print bed to slide across it. These designs are usually very economical but may not be capable of the same quality as designs that have dedicated smooth rods.

Whether it's a rod or a rail, there are a few characteristics that are important for them to work well:

- The smoothness of the surface
- The hardness of the surface
- The straightness of the part
- The stiffness of the part
- The diameter of the part

The smoothness of the surface is important for reducing friction, of course. But the other qualities need a little more explanation.

The importance of surface hardness, for example, may not be immediately obvious. After all, it's not like you'll be hammering it or anything. But what you will be doing is continuously running a bearing back and forth along the length of the part. Over time, the bearing can gouge the surface, making the surface significantly less smooth. And as you know, without a smooth surface, friction can be a problem.

How straight the rod or rail is also important for a reason that should be readily apparent—if it's not straight, movement won't be perfectly linear. But what may not be apparent is exactly how straight it needs to be. In manufacturing, straightness has its own specific geometric tolerance (a specification for how closely the real part must match the design). This tolerance is very tight in rods and rails manufactured for use in linear motion systems but may not be quite so strict for rods and rails that are specifically made for this purpose. For that reason, it's important to be careful about choosing those components when building a 3D printer.

Luckily, the reason rods or rails need to be stiff is straightforward. Rods or rails used in linear motion systems simply can't have any flex in them under the intended loads. This and the other characteristics mean that only a few materials are both suitable for the application and affordable.

Of those materials, the most commonly used seems to be hardened stainless steel. When manufactured properly, this material can possess all of the desired characteristics while also being corrosion resistant and fairly inexpensive. Other materials can be used (and may even be necessary in some industries), but for 3D printers, there doesn't seem to be any reason to use anything else.

The last factor, particularly for smooth rods, is the diameter of the rod. While the rods don't generally provide structural support for the frame, they do need to support the weight of the moving parts. How much weight they need to support and how long they are will determine the diameter needed for the rods. This varies from one 3D printer to another, but smooth rods are usually somewhere between 8mm and 12mm (although they can be bigger or smaller in some cases).

Bearings

As far as mechanical components go, bearings are some of the most fundamental parts in use; in fact, they're almost comparable to fasteners like nuts and bolts. Anytime there are moving parts, bearings practically become a necessity. The purpose of a bearing is simply to reduce friction between moving parts. There are two basic types of bearings: solid bearings (usually called *bushings*) and roller bearings.

A solid bearing doesn't have any rolling elements and is usually meant to be lubricated. The bearing itself is made of either metal or plastic. When it's made of metal, the metal should be of a different type than the shaft it's in contact with. Generally, a softer metal is ideal so the bearing wears instead of the shaft. Combined with a lubricant, a solid metal bearing can be satisfactory for reducing friction.

HOT TIP

When looking at bearings, be aware that many different names are used to describe essentially the same things. For example, solid bearings are often called *sleeve bearings*. This is a result of their long history, their use in many different industries, and very minute differences between similar bearings.

Solid plastic bearings are made from self-lubricating types of plastic like nylon, Delrin (the trade name for a strong, self-lubricating plastic), or some variation of polyethylene (like UHMW, HDPE, or LDPE). These types of plastics have very low coefficients of friction, allowing them to be used for solid bearings without any lubricant. This makes them ideal in mechanical systems where lubricant is impractical, like in sealed systems or when lubricant can damage the mechanisms.

The primary advantage of solid bearings has traditionally been cost. Because there are no small moving parts, they're inexpensive to make, which reduces the overall cost of the machine being built. They were also the first type of bearing used, before the Industrial Revolution made roller bearings feasible. Even now, solid bearings are still the most common type of bearing in use.

But when a mechanism needs to have as little friction as possible, roller bearings are the answer. These types of bearings, often called *ball bearings*, usually have three components: an inner sleeve, an outer sleeve, and a number of small balls between the two. As one sleeve is rotated, it rolls on the balls. For linear ball bearings, the shaft itself acts as the inner sleeve and the balls roll between the shaft and outer sleeve.

Both solid bearings and roller bearings come in two basic styles: linear and rotational. The application determines which kind is needed. If the moving parts only need to rotate on a single axis, rotational bearings are used. If linear motion is needed, like in a 3D printer, linear bearings are used. Therefore, no matter the type of bearing, your printer will need them to be linear.

Stepper Motors

With all of this motion happening, there has to be some kind of electric motor involved to actually make things move. There are many types of electric motors that are technically capable of providing this motion. However, there is one type in particular that is perfectly suited to the task: a stepper motor.

The key characteristic of a stepper motor is its ability to rotate very precisely. A plain old electric motor will simply rotate when electricity is applied, and will rotate faster as the amount of electricity being applied is increased. This simplicity means that a regular electric motor is very efficient, but it also means they're difficult to precisely control. This is what stepper motors were created specifically to address.

A stepper motor is a type of electric motor constructed specifically to allow precise control. That precision is important for producing accurate and detailed prints. It does this by rotating in a series of steps, instead of just rotating continuously when electricity is connected. A common stepper motor, like what is used in most 3D printers, can have 200 steps for every complete rotation of the output shaft.

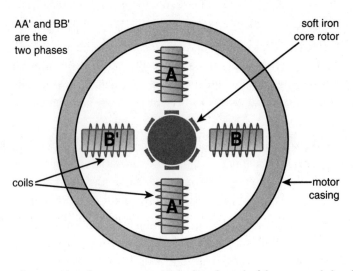

The internal construction of a stepper motor. Notice how the teeth of the rotor are designed so that when they are aligned to one phase, they are misaligned to the other.

> While stepper motors are by far the most common type of motor for 3D printing, there are other options. Continuous-rotation servos can be used and are also fairly accurate. And, in a method similar to what paper printers use, a plain electric motor can be used in combination with an encoder strip to monitor position. But the ease of use and high precision of stepper motors continues to make them ideal for 3D printing.

Because the stepper motor can move a single step at a time, that means it can rotate the output shaft precisely 1.8° for each step (for a 200-step motor). Additionally, a technique called *microstepping* can be used to reduce each step even further. Microstepping is handled by the stepper motor controller, and many controllers are capable of $1/16$ microstepping. That means each individual step can be broken up into 16 microsteps.

At 16 microsteps on a 200-step motor, the output shaft is capable of being positioned at 3,200 individual places per rotation. That's one position every .1125°, an incredible level of precision that simply isn't possible with any other kind of electric motor.

This high level of precision is what makes stepper motors ideal for machines like 3D printers. Even without using microstepping, the accuracy is very high. If a full rotation of the stepper motor output shaft moves an axis 1mm, each step will move it .005mm. That means each axis of the 3D printer can be positioned anywhere within a tolerance of +/- .0025mm, and that can be made significantly more accurate with microstepping.

With that kind of accuracy, it's easy to see why stepper motors are so popular, and why they're used in virtually all 3D printers. But how does the rotational movement of a stepper motor translate into the linear motion needed for 3D printing? That's where belts, pulleys, and lead screws or threaded rods come in.

Belts and Pulleys

Belt and pulley systems are one of the oldest mechanical systems in history and have a wide range of uses. Aside from simply transmitting motion from one pulley to another, they can be used to reduce a drive system to gain *mechanical advantage*.

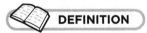

> **Mechanical advantage** is the amplification of force with the use of a tool or mechanical system. This is usually achieved by trading movement distance for force. A lever is the most basic example of this, because if one side of the fulcrum (pivot point) is twice as long as the other, it will double the force exerted (though it will also double the distance it needs to be pushed). This same basic concept is applied in a vast array of machines using things like gears, pulleys, screws, and so on.

In its most basic setup, a belt and pulley system is very simple. One pulley is driven by a motor of some kind, and a belt connects it to a second pulley to transmit the rotation of the first pulley to the second. If both pulleys are the same size, the drive system has a 1:1 ratio and the purpose of the system is simply to transfer power.

Mechanical advantage is achieved when the pulleys are of different sizes. If the first pulley (the drive pulley) is smaller than the second, the resulting output has more torque but less speed. When the system is reversed and the first pulley larger than the second, the output has greater speed but reduced torque.

Multiple pulleys can be used to gain even greater reductions and significant mechanical advantage. Systems like that are what allow a person to lift loads many times their own weight. In the same way, such a system can allow a relatively low-torque (but high-speed) motor, like an electric motor, to handle high loads.

However, while some printers use mechanical advantage in their designs (especially for the cold end), others keep power transmission at a 1:1 ratio. To move the print bed, for example, a pulley is attached to the output shaft of a stepper motor and a second pulley is mounted on the opposite end of the axis. A belt is looped around them and attached to the print bed in the middle. As the output shaft of the stepper motor spins, it pulls the belt and therefore moves the print bed along the axis.

As far as pulley systems go, this is a very simple setup. However, mechanical advantage can be introduced by changing the size of the drive pulley. The bigger the drive pulley is, the further it will move the print bed per revolution. If the speed at which the stepper motor turned remained constant, the size of the drive pulley would determine the speed and torque of the print bed movement. A large drive pulley would move it quickly but with reduced precision and torque. A small drive pulley would move it slowly but with higher precision and more torque. This allows the printer manufacturer to prioritize speed or precision in their designs.

While the setup is pretty much standard, the physical design of the belts and pulleys on 3D printers can vary. They can be as low tech as a fishing line acting as a belt, wrapped around a drum acting as pulley. Or they can be high-tech parts designed specifically for this purpose. Most commonly, though, 3D printer manufacturers use simple toothed belts and pulleys.

These toothed belts and pulleys are fairly similar to a car engine's timing belt. It's a rubber belt with teeth on one side that mesh with pulleys that also have teeth. This setup is ideal because it doesn't slip, the belt doesn't stretch, and it's affordable.

Belt and pulley systems are great for horizontal movement (the X and Y axes), where the belt doesn't have to bear any weight. All it has to do is pull one way, and then back the other way.

This is a common GT2 belt and pulley setup.

Weight-Bearing Components for Converting Motion

When rotational motion (for the vertical Z axis) needs to be converted into linear motion, and there is a significant load on the system, lead screws or threaded rods (which are inexpensive alternatives to lead screws) are the best option. There are two reasons lead screws are ideal for such a situation: they can provide a significant mechanical advantage without complicated pulley or gear systems, and the lead screw itself can bear the weight of the components it's lifting.

A lead screw drive system is a lot like a bolt and nut; in fact, that's pretty much what it is (physically speaking). When you rotate the bolt but don't allow the nut to spin with it, it pushes or pulls the nut along the rotational axis. A lead screw works in the same way: the stepper motor turns the lead screw, which pushes or pulls the nut up and down on the Z axis. The nut is attached to whatever components need to be moved in the Z axis, allowing the stepper motor to lift or lower them as it turns.

Lead Screws

Lead screws come in all kinds of sizes to suit the requirements of the machine being built. They come in various diameters and *pitches* (metric and standard) with one start threads, two start threads, or even more. The diameter is factor in the strength and stiffness of the lead screw and can help the smooth rods constrain movement to a single axis. The thread pitch determines the mechanical advantage of the drive, because it is what controls how far the nut is moved per rotation of the lead screw.

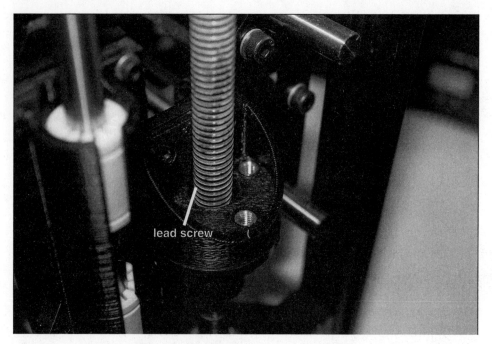

A lead screw and nut linear motion system for the Z axis.

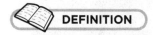

> The **pitch** of a thread is the distance from one thread to the next. This is what determines how far the screw and nut will move (relative to each) with one full rotation. For example, an M8 screw has a standard pitch of 1.25mm. So if you have a lead screw with an M8 thread, every full rotation of the lead screw will move the nut 1.25mm.

Despite how common and utilitarian lead screws are, they are still very precisely made parts. The complexity of the specifications of lead screws (they need to be smooth, straight, hard, and stiff), along with the precision required when machining them, makes them very expensive.

Many 3D printer designs require two of them (one on either side of the printer), which just adds to the bill. In fact, this cost is so high that lead screws can end up being the most expensive parts on a consumer 3D printer.

With such a high price for such a basic part, 3D printer manufacturers and DIY builders have tried to find another solution.

Threaded Rods

In an effort to reduce the cost of 3D printers, many designers have chosen to substitute threaded rods for lead screws. A threaded rod is essentially a very long headless bolt. The general purpose of a threaded rod is for fastening, which is done by threading a nut onto either end.

Because threaded rods are only intended to be fasteners, the manufacturers put little importance on the qualities that are considered necessary for lead screws. The result is a part that isn't particularly straight or hard and doesn't have perfectly machined threads. A threaded rod is just good enough to thread a nut onto while being sort of straight.

Manufacturers are able to make threaded rods at very lows costs while still maintaining the required precision. That cost is so low that threaded rods are generally about $1/10$ the price of lead screws. At that kind of price, it's easy to see their allure to a 3D printer manufacturer trying to build an affordable consumer printer.

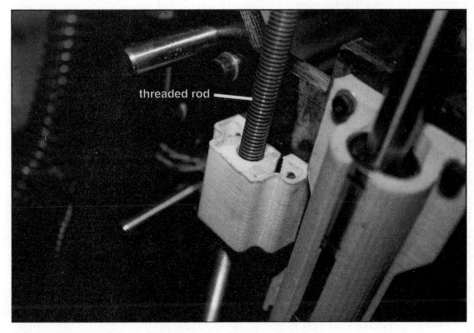

A threaded rod used in place of a lead screw.

But there is a pretty glaring problem here: if threaded rods work, why would anyone spend the extra money on lead screws? From the point of view of those who use threaded roads, there are a couple of different answers to that question.

One possibility is that lead screws are necessary for certain kinds of machines but not for 3D printers. The argument is that the loads being put on the lead screw by a 3D printer are small, and the precision required is fairly low. If a real lead screw makes a difference at all, it's very subtle.

Another possibility is that there is a perceptible difference between lead screws and threaded rods but that it can be designed around. The idea here is that lead screws are only necessary if they're firmly constrained, in which case they're acting kind of like a smooth rod to restrict movement perpendicular to the Z axis. But because there are already smooth rods for this purpose, it is unnecessary for the lead screw to provide that function. Because the threaded rod doesn't have to restrict perpendicular movement, it's not necessary for it be perfectly straight. All that matters is that it doesn't inadvertently introduce perpendicular movement.

At this point, there is very little evidence for or against the effectiveness of threaded rods that isn't just anecdotal. The quality of a 3D-printed part is the result of the complex interaction of many different pieces of hardware and many settings in software. It's difficult to control every other factor while only changing from lead screws to threads in order to scientifically test this. However, the anecdotal evidence seems to suggest that threaded rods can work well if the design is done properly.

Attachment and Connection Components

Whether lead screws or threaded rods are being used, there always needs to be a way to connect the stepper motors to them. And, once that stepper motor is spinning, there needs to be a way to translate the turning screw into movement on the Z axis. Both of these problems have specific parts designed to solve them: a coupler connects the stepper shaft to the screw, while a nut attaches parts to that screw so they can be moved along the Z axis.

Couplers

The job of the coupler is rather unspectacular: it's just there to connect one shaft to another. However, various kinds of couplers exist to serve all sorts of specific purposes. There are couplers designed to flex and others that are designed specifically to minimize flex. Some couplers are supposed to be springy to absorb shock in the shaft, while others are designed to avoid this.

In the case of 3D printers, which type of coupler is ideal depends on whether lead screws or threaded rods are being used. If a lead screw is being used, it's reasonable to assume it is both very straight and very stiff. If that's the case, the coupler should be solid and stiff as well, so that movement is smooth and constrained to that rotational axis.

 FASCINATING FACT

In many RepRap 3D printer designs, the coupler is simply a small piece of flexible plastic tubing that fits over the stepper motor shaft and the threaded rod. The tubing is then zip tied on both ends to keep it tight. This low-tech solution is both very inexpensive and surprisingly effective.

However, if a threaded rod is being used, it's very likely it's not straight at all. It's probably slightly bent and maybe a little bit flexible itself. In such a case, it's preferable for the coupler to be flexible in order to absorb the movement caused by the threaded rod not being straight.

A standard coupler meant to allow some flexibility.

The coupler itself can be made from a variety of materials and in a variety of styles. Aluminum, steel, and brass are all popular materials, and types of plastic can be used as well. Designs can be a simple straight tube, can be cut in a helix to absorb forces along the rotational axis, or can be cut with slots to absorb forces perpendicular to the rotational axis. Most will have some sort of set screw or clamping apparatus to lock the coupler onto both shafts, but even that may change in some specific circumstances.

Nuts

A nut is a simple part that is threaded onto a threaded rod or lead screw. It can be a basic hex nut like the kind you curse whenever you have to do work on your car, or it can be a specialized type of nut meant specifically for linear motion systems.

Like most other components, materials can vary. However, it's widely considered to be necessary for the nut material to be softer than the material of the lead screw or threaded rod. The reason is basically the same as why bearings and bushings are made of softer metals: so that they wear first. As expensive as lead screws are, it would be a real shame if the threads got worn down by a 50-cent nut. It's much better to replace the inexpensive nut as it gets worn down.

With that said, nuts that are used on 3D printers aren't often as simple as the kind you buy in bulk at the hardware store. Because these nuts are meant for power transmission, not fastening, they have two unique qualities:

- They need to have some kind of mount on them.
- They should resist backlash.

The need for a mount is pretty straightforward; there has to be way to attach the parts you want to move. For this reason, there are nuts made with mounting flanges so the moving parts can be connected. These mounting flanges come in various sizes, shapes, and hole patterns, but those details aren't really relevant to their operation.

What is relevant to the operation of a screw drive system is backlash. Backlash, in this context, is when the nut slightly shifts when the rotation of the screw reverses. The direction of movement suddenly changes and the nut moves slightly because of the space between the threads. To be clear, this is a very small shift, comparable to the movement you feel if you try to wiggle a nut that's threaded onto a bolt. But I'm talking about a machine that is trying to move with very high accuracy, so even a slight shift can affect that.

The solution to this problem is the creatively named *anti-backlash nut*. A very basic design for this consists of four parts: a housing, two normal nuts, and a spring. The housing fits over the nuts and keeps them from moving independently of each other, while the spring sits compressed between the two nuts. Because the spring is compressed, it's constantly pushing out on both nuts. This ensures that the top surface of one nut's threads is in contact with the screw, while the bottom surface of the other nut's threads is in contact. This effectively eliminates backlash, because when the direction is changed, one of the nut's threads will always already be in contact with the screw.

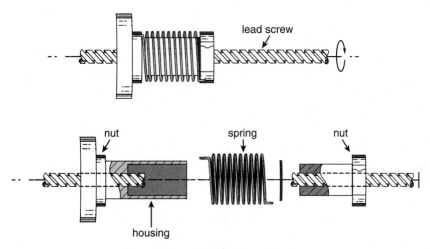

A simple anti-backlash nut design.

With all of this detail on the nuts and screws of 3D printers, it may seem like they're very complex machines. And in a way, they are. But the reality is that most machines are at least this complicated when you really get into the details, and 3D printers aren't an exception.

The Least You Need to Know

- Smooth rods, rails, and bearings are used to provide smooth, low-friction movement.
- Stepper motors are unique types of electric motors that are perfectly suited for 3D printing because of their precision.
- Belts and pulleys are often used to drive the X and Y axes of 3D printers, while lead screws are used to drive the Z axis.
- Threaded rods can be used in place of lead screws, but there is some ambiguity about how well they do the job.
- Couplers are used to attach lead screws or threaded rods to stepper motor shafts, while nuts (anti-backlash or otherwise) are used to attach the moving parts to the lead screws.

CHAPTER 7

The Extruder

Extrusion is the key to FFF (and FDM) 3D printing. It's what separates the FFF printing process that is so common in consumer printers from the other processes that are used in more expensive 3D printers. Extrusion itself is used in many ways in manufacturing, but the technique has been fine-tuned for 3D printing in FFF printers. In this chapter, I cover how extrusion works and how it affects your prints.

In This Chapter

- How FFF extrusion works
- Different types of cold ends and hot ends
- Changing nozzles
- When to use a print fan
- The benefits of multiple extruders

What Is Extrusion?

In general terms, extrusion is any process used to eject material with a specified cross-section. Extrusion in manufacturing can be used to create a wide range of parts, with the major restriction being that the cross-section of the part has to remain constant.

With FDM/FFF 3D printing (see Chapter 2), extrusion is the process used to actually deposit new plastic onto the print bed. It also refers to how the aluminum frame pieces that are often used to build 3D printers are created.

Complex shapes are possible with extrusion in manufacturing, but the process is much simpler on 3D printers. Plastic is extruded out of a plain circular nozzle, resulting in a very simple cylindrical extrusion.

To extrude plastic, FFF printers use a system known as an *extruder* that feeds filament into a hot end to be melted, and then deposits it on the print bed. How exactly this is done varies a little bit depending on the particular 3D printer, but they're actually all pretty similar.

The Cold End

The system begins with feeding the filament, a task handled by a mechanism called a *cold end*. The cold end pulls filament and pushes it into the hot end. As more filament is pushed into the hot end, the melted filament is squeezed out through the nozzle for deposition onto the print bed. So the purpose of the cold end is easy to understand; it's just there to feed the filament. But it's more than just that; it needs to be able to feed the filament precisely and consistently.

HOT TIP

The terminology used for various parts of 3D printers can often be a bit confusing. Sometimes, different terms are used to describe the same part based on the manufacturer, plus the hobbyist nature of 3D printing has led to parts being developed simultaneously by independent designers.

The cold end is one example of this confusion. The cold end is what feeds filament into the hot end, which in turn melts the filament. However, the cold end is often referred to as the extruder instead. This is made even more confusing, because *extruder* is the term used to describe the entire system (cold end and hot end). To try and avoid any confusion in this book, I'll refer to the individual mechanisms as the cold end and the hot end, and the system as a whole as the extruder. But be aware that these terms may be used differently depending on the source.

It's important that the printer be able to extrude exactly the right amount of filament in order for the print quality to be good. As I mentioned in Chapter 3, the layer thickness is determined in part by how much filament is being extruded. The slicing software knows exactly how much filament is being pushed out of the hot end at any given time and calculates the extrusion as needed.

In order for the slicer to be able to make usable calculations, the cold end has to be capable of consistently feeding exactly the predicted amount of filament at all times. If the filament is fed inconsistently, the calculations being made by the slicer software become useless, resulting in either overextrusion or underextrusion (both of which will yield a poor-quality print).

For this reason, the extruder has to be properly calibrated. Extruder calibration is absolutely necessary to make sure the correct amount of filament is being extruded. This is a simple matter of making sure the amount of filament being fed into the hot end matches the predicated amount and adjusting the feed rate if it's not right.

To do this, cold ends use stepper motors for precise control—the same type of stepper motors that are used for moving the axes of the 3D printer. The stepper motor turns some sort of hobbled bolt or drive gear that has serrations to grip the filament, and a bearing on a spring provides tension. The filament is held tight between the bearing and hobbled bolt, and as the stepper motor turns, it pushes the filament into the hot end.

The setup for this mechanism depends on what kind of cold end is being designed, and there are two distinct types: direct feed and Bowden.

Direct Feed

The most common type of cold end is the direct feed style (not to be confused with direct drive, which I discuss later in this chapter). It's called a *direct feed* cold end because the hot end is located right below it, and filament is fed directly from the drive gear into the hot end.

There are two advantages to this direct feed style: it's a simple setup, and it reduces the chances of the filament binding or bunching before it enters the hot end.

The simplicity of the setup is probably the primary reason why direct feed cold ends are so popular and why almost all consumer 3D printers use them. The motor and drive gear are mounted right above the hot end, and the filament is pulled from a spool, which can be mounted anywhere on or around the printer.

A direct feed cold end made by Printrbot.

But as consumer 3D printer technology has matured and our understanding of it has improved, it has become apparent that the real benefit of a direct feed cold end is that it lessens the likelihood of the filament bunching. As flexible filament (a material which feels similar to rubber or silicone) and 1.75mm filament have gained popularity, it's become more of a problem for filament to bunch up and jam before it actually enters the hot end. Having the drive wheel right above the hot end to push the filament doesn't leave any space for it to bunch up.

Bowden

Bowden-style cold ends are named after a mechanism called a *Bowden cable*. A Bowden cable is a push/pull device that consists of a hollow tube with a solid or stranded cable running through it that lets you transmit pushing or pulling force over a distance with a flexible cable. This mechanism is used in all sorts of things, including airplane controls, bicycle brakes and gear shifters, motorcycle clutches, throttle cables, and many other things.

> **FASCINATING FACT**
>
> There is some debate over who Bowden cables are named for. The original Bowden mechanism was invented and patented by Ernest Monnington Bowden, but Sir Frank Bowden (founder of the Raleigh Bicycle Company) is often given credit for its invention. The two men were not related, but their common last name is surely a factor in the confusion. Ernest Bowden actually licensed his patent to Sir Frank Bowden for use in his bicycle brake systems, which just further confuses things.

A Bowden cold end works on a similar principle. Instead of having the drive mechanism right above the hot end on the extruder carriage, it's located elsewhere on the 3D printer. A hollow and flexible tube (usually made from a low-friction plastic like polytetrafluoroethylene [PTFE]) runs from the cold end to the hot end, and the filament is pushed through it. This setup serves one very specific purpose: to remove mass from the extruder carriage.

A Bowden cold end on an Ultimaker 2.
(Courtesy of Ultimaker)

Having a smaller and lighter extruder carriage is advantageous for two reasons:

- It reduces the overall size of the 3D printer.
- It improves print quality.

It reduces the size of the printer because the extruder carriage doesn't need as much room to reach the ends of the print bed. How it affects print quality is a little more complicated.

Moving a large amount of mass around quickly is a difficult task. Every time the direction of the moving mass changes, the printer has to overcome the inertia of that mass. But unless you're printing at slow speeds, the 3D printer never completely overcomes the inertia and it slightly overshoots the intended stopping point. When it overshoots, it results in uneven edges around features, making fine details impossible to print well.

A large, moving mass also creates vibrations as the extruder carriage moves back and forth. This results in an effect called *ghosting,* where faint outlines of a feature show up on the surfaces near those features. So the mass of the extruder carriage ends up having a pretty significant effect on the output quality of a 3D printer.

Bowden style cold ends solve this problem by removing as much mass as possible from the extruder carriage. It takes the weight of the stepper motor and drive gear and moves it to a stationary location. But, of course, this setup does have its drawbacks.

In the previous section, I explained how direct feed cold ends reduce the chances of filament bunching before it enters the hot end. And that's exactly the problem with Bowden cold ends. With such a long distance between the drive gear and the hot end, there is a lot of room for the filament to bunch and jam. This makes it difficult to use 1.75mm filament and almost impossible to use flexible 1.75mm filament.

Direct Drive vs. Geared

Whether a Bowden or direct feed cold end is being used, there is still the option of pushing the filament either with a drive gear connected directly to the stepper motor shaft or with a hobbled bolt connected to a gear system. As I'm sure you're sick of hearing by now in this book, each has its advantages and disadvantages.

Direct drive extruders use a drive gear that is mounted onto the stepper motor shaft. This is the absolute simplest way to design an extruder, which is always a good thing. However, it takes a lot of force to push filament through a hot end. Because they are light and compact, direct drive extruders have difficulty handling the force needed for 3mm filament.

On the other hand, geared extruders use a pair of gears to gain mechanical advantage and the torque needed to force 3mm filament through the hot end without difficulty. The disadvantage, of course, is that the gears add size and weight to the extruder carriage, although that's not a problem for Bowden extruders since the cold end doesn't move.

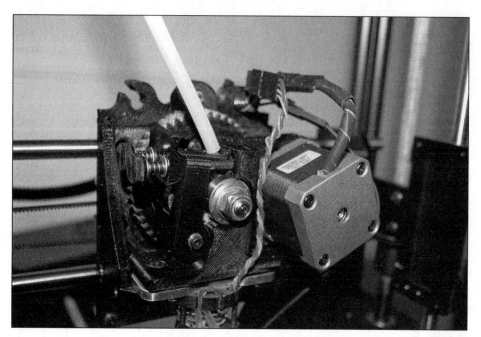

A Greg's Wade Reloaded geared direct feed cold end.

The Hot End

As you've learned in previous chapters of this book, the hot end serves one very specific purpose: to melt the plastic filament. It must melt the filament very quickly in order for the 3D printer to print at a reasonable speed. For that to happen, the temperature of the hot end has to be very high.

Exactly how high the temperature needs to be depends on the material being printed, the filament manufacturer, and even sometimes the specific batch of filament. For the most common materials, however, the hot end needs to be operating at temperatures between 180°C and 250°C. To put that into perspective, that's approximately the same temperature as the hot oil in a deep fryer.

An Ubis hot end, which is used on Printrbot 3D printers.

Some materials, however, require much hotter temperatures. For example, polycarbonate is an extremely tough plastic that requires temperatures of 300°C or more to print well. With such high temperatures, it's easy to see that some special hardware is required. In order to heat up the hot end, two specific electronic components are needed: a thermistor and a heating element.

Thermistor

A thermistor is a type of variable resistor similar to a *thermocouple* that changes based on how hot it is. The term thermocouple is often used incorrectly to describe the thermistor used in 3D printers, but they're actually entirely different components. They work similarly, but thermistors are generally more accurate and are better suited for 3D printers.

The thermistor is mounted onto the hot end at the point where the filament is supposed to be melted. As the hot end heats up, the resistance of the thermistor changes. The control board of the 3D printer monitors this resistance change and uses it to calculate the temperature of the hot end. In essence, it's basically just a very accurate thermometer that can be read by the 3D printer.

> **DEFINITION**
>
> A **thermocouple** is a type of temperature-measuring device used in a wide range of industries. It's inexpensive and doesn't require a power source, which makes it ideal for some applications. However, thermocouples aren't very accurate, which generally makes them unsuitable for use in 3D printer hot ends.

Thermistors are accurate to within a fraction of a degree when properly set up. Because thermistors are made by a variety of manufacturers, sometimes for slightly different applications, the proper setup can vary. To solve this problem, 3D printer firmware comes with information for many different thermistor models in order to make the temperature reading as accurate as possible. With an accurate temperature reading, the 3D printer can precisely control how hot the hot end is. This allows the user to set the temperature as necessary for a specific type of filament.

But getting any accurate temperature reading is only half of the equation; the 3D printer also needs to be able heat up the hot end to the desired temperature using a heating element.

Heating Element

The heating element for a hot end is a much more general type of component than the thermistor. Virtually all electronic heating elements work in the same way, whether it's for an electric space heater, an electric stove, or the hot end of a 3D printer. All of these devices use a method called *resistive heating*.

Resistive heating is an extremely basic process, and is really just a fundamental property of electronics in general. When current passes through a conductor, it heats the conductor. If the voltage in a circuit remains constant, the heat of a resistor increases as the resistance decreases. The current draw also increases along with the heat. So when the objective is to heat something up, like a hot end, a heating element with a low resistance is used.

That low resistance draws a high current, and therefore a lot of power needs to be dissipated by the source of the resistance (in this case, the heating element). That power is dissipated in the form of heating up the heating element, and in turn the hot end.

Physical Design and Makeup

The combination of the thermistor and the heating element allows the 3D printer to set the temperature of the hot end very accurately. When the 3D printer is first turned on and the hot end is still cool, constant voltage is applied to the heating element to heat it up as fast as possible. As it gets close to the desired temperature, that voltage is pulsed to slow down the heating process until it reaches the correct temperature.

Once it's at the correct temperature, the printer monitors the thermistor and turns the heating element on and off as necessary to maintain the temperature. It does this with a method called *pulse-width modulation (PWM)*, which flips the voltage on and off very quickly to average out into a specific voltage to keep the temperature constant.

So how is it designed to make this process happen? The design of a hot end is a reflection of a few goals:

Maintain a constant temperature near the nozzle to melt the filament. This is achieved with a heating block where the thermistor and heating element are located. The block is heated and monitored, and has a hole running through where the nozzle is mounted. As the filament enters the heating block, it's rapidly melted into molten plastic to be squeezed through the nozzle.

Heat up the heating block quickly. The heating block is generally made from a metal with a high thermal conductivity. This allows the heat from the heating element to be rapidly transferred to the heating block; therefore, it heats up quickly.

Get cool over a very short area so it can be mounted to the 3D printer. The mount is often made of the same type of plastic that is being extruded, so the hot end needs to be cool at that point in order to not melt the mount itself. To make that possible, a heat sink is often located between the heating block and the mount. In some designs, there is a fan located on the heat sink to cool it even more. This can lower the temperature from a very hot 300°C to roughly room temperature over just a couple of inches.

HOT TIP

Hot ends are often interchangeable, depending on the type of mount they use. Progress is rapidly being made in improving hot end designs, and new models are frequently being released. So it's probably not necessary to choose a 3D printer based on the hot end as long as it uses a common or replaceable mount.

Traditional hot end designs incorporate a variety of materials, including some high-temperature plastics. For common filament materials like polylactic acid (PLA) and acrylonitrile butadiene styrene (ABS), this is perfectly acceptable because the plastic can handle the temperatures required for those materials. But for some materials that have higher melting points (like polycarbonate), the hot end has to be heated to temperatures that the plastic in the hot end can't withstand.

For that reason, all-metal hot ends are becoming popular. A hot end made completely from metal, instead of using some plastic parts, can be heated to much higher temperatures without risking damage to the hot end itself. With an all-metal hot end, many more material types can be printed successfully, allowing the user a wider range of possibilities.

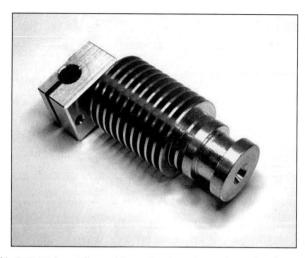

This E3D V5 is an all-metal hot end and requires active cooling from a fan.

But making a hot end completely out of metal presents some design challenges that are still being overcome. One of the biggest problems is that the heat from the heating block is transferred more efficiently across the entire hot end, meaning the mounting point can get too hot for the mount. To solve this problem, much larger heat sinks are needed, and a fan is often required. But the versatility of all-metal hot ends can definitely make the challenges worth the trouble.

The Nozzle

The nozzle is often considered to be part of the hot end, but it's a replaceable part that I think deserves its own section. Hot ends almost always come with a compatible nozzle, but hot end manufacturers also generally sell additional nozzles in a variety of sizes.

Why would you want to purchase different-size nozzles? Because, as I've explained in Chapter 3, the size of the nozzle is one of the factors that determines layer height—the larger the nozzle, the more of an increase in layer height. Nozzle size isn't the only thing that determines layer height, so there is some wiggle room for each nozzle size. But if you want to print a layer height that is significantly different, it can be worth changing nozzles.

Nozzles that come with hot ends are usually somewhere between .25mm and .50mm, because that's a general size that is useful for most printing. But nozzles can be much smaller for specialized printing where very fine detail is required. Alternatively, some prints can be done with larger nozzles in order to print large objects more quickly.

WATCH OUT!
Nozzles aren't generally compatible between different hot end models. This is because hot ends usually have nozzles that are made specifically for them. They use different threads, are different sizes, and are sometimes even different shapes. For this reason, it may be a good idea to check ahead of time what nozzles are available if you know you're going to need a smaller- or larger-than-usual nozzle.

Print Fans

In addition to the hot end fans that are used to cool the hot end for mounting, one or more fans can also be used to cool the printed object itself. The fan is directed toward the extruded filament as it exits the nozzle in order to quickly cool the plastic.

The purpose of cooling the plastic is to get from the high temperature needed for extrusion to room temperature as quickly as possible. Plastic expands when it gets hot and contracts as it cools. That contraction can result in warped parts, because the entire part cools and contracts together. So a fan can be used to cool the filament quickly to avoid the entire part contracting at once, which helps to reduce warping.

Some filament materials, like PLA, also require active cooling from a fan to combat deformation. PLA is very soft until it cools, and the plastic can sag as its being printed if it's not cool. This is especially apparent when printing overhangs or bridges, which will sag dramatically if the plastic is too hot.

Having a fan blowing on the plastic coming out of the nozzle allows you to print overhangs and bridges without them becoming deformed. Just a fan blowing in the direction of the nozzle can do a lot to help, but in order to cool effectively, fan shrouds can be used. The shroud acts as a sort of funnel to direct air to a very small area just below the nozzle where the filament is being deposited. This very quickly cools the filament as soon as it exits the nozzle, so the plastic gets hard and doesn't deform.

Whether you should use a fan is determined by the material you're using. For PLA, a print fan is virtually a necessity. For other materials, like ABS, a fan can actually be detrimental to print quality. ABS isn't as soft as PLA when it's extruded, so a fan isn't needed to keep it from deforming. In fact, actively cooling ABS with a fan can make it solidify too much and can keep it from sticking to the previous layer.

If the 3D printer is equipped with a print fan, it's usually controlled via software. The slicer software can turn the fan on and off (or even on at different speeds) throughout the printing process. It's common to have the fan on at different parts of the print (like for overhangs) and off for others (like the first few layers). In later chapters, I'll be going over when and how to use print fans, but for now, you should just be aware of its purpose.

Using Multiple Extruders

All FFF 3D printers have at least one extruder, but it's also possible to have multiple extruders on the same printer. These are mounted right next to each other so the printer can use all of the extruders at the same time. The most basic reason for doing this is so you can print a single part in more than one color. But the possibilities introduced by using dual extruders (or more) can be much more exciting than just making colorful parts.

The Kraken, also made by E3D, has four extruders built into a single unit.
(Courtesy of E3D)

For Support Material

One very useful way to take advantage of a dual extruder setup is to use one extruder for regular filament and the other for support material. There are special kinds of filament which are formulated specifically for this task. Some manufacturers, such as Stratasys, have developed their own formulas for accomplishing this task, but in many cases patents keep these from being available on printers from other manufacturers. When it comes to the most common support material in consumer printing, people tend to use high-impact polystyrene (HIPS). HIPS is a type of plastic that dissolves in a common solvent called *limonene*. Limonene doesn't affect PLA or ABS, so HIPS can be used for support material and then dissolved away, leaving the printed part untouched.

Using a soluble support material like this in the second extruder eliminates the hassle of having to manually remove support material made from the same material as the part. Not only is it a lot of work to remove support material, it can be difficult to do without damaging the part if there are fine details. In some cases, support material could be almost impossible to remove if, for instance, it's supporting internal geometry that can't be reached.

Dual extruders with a standard material and soluble support material completely solve this problem. You can just print the part with the two types of filament, and when the print is finished, you can soak it in a limonene solution. After a few hours, the support material is dissolved away and your finished part can be removed.

For Filaments with Different Properties

Another use for dual extruders is printing two kinds of filament with different properties at the same time. A single part could be printed with a hard ABS plastic and also a flexible filament, for example. You could print a plastic tool with a soft rubbery handle. Or a part could be made with flexible joints to allow movement.

With the wide range of filaments on the market today, there are a lot of possibilities for an imaginative person. You can take advantage of the different melting points, coefficients of friction, hardness, and flexibility of various filament types to create unique objects. This is an area of 3D printing that is just starting to be explored, and a lot of unconventional designs are possible.

HOT TIP

It doesn't stop with just dual extruders either. You can already purchase a four-extruder assembly meant for consumer printers that lets you print four different filament types simultaneously. You could use this to print two different colors of ABS, a flexible filament, and HIPS support material all at the same time. Or you could use it for any other combination of filament materials that you think would be useful. Imagine the kinds of designs you could come up with!

The Least You Need to Know

- An FFF extruder consists of a cold end that feeds filament into the hot end.
- Direct feed cold ends feed more reliably, while the lower mass of Bowden cold ends can improve print quality.
- In order to heat up the hot end, two specific electronic components are needed: a thermistor and a heating element.
- Multiple extruders can add a great deal of versatility to the kinds of prints a 3D printer is capable of.

CHAPTER 8

The Build Platform

When it comes to 3D printers, every part on the machine is important to its functionality. That's true even for things that seem simple and uncomplicated, like the build platform (also called a *print bed* or *build plate*). The build platform is actually just as crucial as any of the other parts and can be just as complicated.

The build platform is where the printer actually deposits the plastic to make the part. This seems straightforward, but in order for the print to be successful, the plastic needs to stick well to the platform. There are a few factors which determine how well the plastic will stick: the flatness of the platform, how level it is, what material the build platform is made of, and any surface treatments which are applied to the platform.

In this chapter, I go through all you need to know about the build platform, including what it's made of, the various parts that make it up, and the surface treatments used on it to improve adhesion.

In This Chapter

- Materials commonly used for build platforms
- The purpose of heated beds and build chambers
- Different surface treatments to improve adhesion

Build Platform Materials

There are a handful of materials that are commonly used for build platforms, such as glass, *borosilicate*, aluminum, and copper. What all of these materials have in common is they can be made very flat. The flatness of the build platform is essential for ensuring the plastic sticks to the bed. This is because the nozzle has to remain at a very specific height above the build platform for the extruded filament to adhere to the bed, and that height wouldn't remain constant if the platform weren't flat.

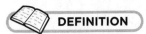

> **Borosilicate,** also known as *Pyrex* (one of its trade names), is a type of glass that is formulated to reduce thermal expansion. Because it's much less prone to thermal expansion than normal glass, it's at a much lower risk of thermal shock. Thermal shock can crack traditional glass if it's quickly heated or cooled unevenly, a problem that borosilicate doesn't experience.

In addition to being flat, the build platform has to be constructed from a material that the molten filament can actually stick to. Some materials are too slick, and the extruded filament simply won't stick to it. An ideal material would let the filament stick very well while it's printing, and then allow it to come off easily when the print is done. If it sticks too well after the print is done, it can be difficult to remove the part without damaging it.

For this reason, specialty build platforms are being developed just for 3D printers. This is a brand new type of product, but the selection should increase dramatically in the coming years. The platforms themselves are made from a variety of materials, depending on the particular manufacturer. But the goal is the same for all of them: for the plastic to stick well while printing and come off easily when the print is done. But in many cases, in order for that to work, the build platform needs to be hot. This is where heated beds come in.

Heated Beds

Heated beds, as the name implies, keep material warm. A heated bed works pretty similarly to a hot end in principle: a thermistor measures temperature, while resistive heating is used to get the bed hot. In this case, though, the entire bed is a big resistor that heats up instead of just a small heating element.

Conventional heated beds are large printed circuit boards (PCBs) specifically designed with large copper traces that act as a very low-ohm resistor (due to Ohm's law). When voltage is applied, this draws a very large amount of current.

A large 12×12-inch PCB used for heating the build platform.

 FASCINATING FACT

Recently, other types of heated beds have started to gain popularity in the consumer 3D printing industry. This includes silicone heating pads, which have long been popular in other industries. Unlike traditional PCB heat beds, these are flexible and are manufactured in a wide range of sizes and shapes.

With all of that power flowing through what is basically a resistor, the heated bed gets hot—anywhere from 60°C to 120°C is common. This, in turn, heats up the build platform. And just like with the hot end, the 3D printer monitors the temperature using the thermistor (see Chapter 7) and adjusts the voltage going to the heated bed to keep the temperature steady.

One of the benefits of a heated bed is reduction in the warping of a part. As I covered in earlier chapters, warping happens when the part contracts as the plastic cools. A heated bed can help combat this problem by keeping the part warm as it is being printed. For some materials, like ABS, a heated bed is basically a requirement. Other materials, such as PLA, don't need a heated bed but can still benefit from one.

This is the reason why some 3D printers include a heated bed and others don't. Generally, only inexpensive 3D printers come without a heated bed. Those printers can usually only successfully print PLA. In order to print ABS (and many other materials), a heated bed is a necessity.

In addition to reducing warping, heated beds also help filament stick to the build platform. Heating the build platform gives the filament a nice, hot surface to stick onto, which is true for all of the material types on the market. While it's not absolutely necessary for all materials, a heated bed does a lot for getting the first layer of the print to adhere well to the build platform.

Heated Build Chambers and Enclosures

A heated build chamber is used to prevent warping. It contains the heat generated by the hot end and heat bed, keeping the air warm and reducing drafts. Unlike an enclosure, a heated build chamber includes active heating. This means a heating element and some kind of temperature sensor (a thermistor, for example) are placed inside of the enclosure to heat it up, almost like an electric oven. This can get the air inside very hot and can almost completely eliminate warping with materials like ABS.

Unfortunately, not all 3D printers include heated build chambers. The main reason is that Stratasys owns the patent for heated build chambers, meaning they use them extensively on their professional FFF 3D printers but other manufacturers can't add them to their printers. If it weren't for that patent, you'd surely see heated build chambers on a lot of consumer 3D printers.

 HOT TIP

Stratasys now owns MakerBot, a popular consumer 3D printer manufacturer. Because Stratasys has the patent for heated build chambers, it's likely that in the future the technology could be used on consumer 3D printers made by MakerBot. If a heated build chamber is important to you, it might be worth waiting for this to happen.

But that patent doesn't keep manufacturers from adding an unheated enclosure to their printers. Many consumer 3D printers have enclosures that can hold in the heat generated by the hot end and heat bed, keeping the air warm and reducing drafts. Warm air around the printer keeps the part warm, so even without active heating, this can do a lot to reduce the warping that is common when printing with ABS.

Surface Treatments

Another way to reduce warping is to make sure the first layer of the print sticks very well to the build platform. Most of the materials that build platforms are commonly made from don't work very well on their own. Even with a heated bed, the filament doesn't always adhere well. So to make sure that plastic really sticks, a variety of surface treatments can be applied to the build platform. These treatments are usually some sort of tape, film, or glue.

The 3D printing community has come up with a lot of novel ways to ensure the filament adheres well to the build platform. A handful of these methods stand out as being especially effective, and have come to be a sort of standard operating procedure for hobbyists doing 3D printing. There are too many to cover all of them, but I'll go over some of the more popular ones here.

Painter's Tape

One of the most popular surface treatments is painter's tape. This is just the plain old blue tape you put around your trim when you paint your walls. The tape is applied with the adhesive on the build platform, so the entire build platform is completely covered. You then print directly onto the nonadhesive side of the painter's tape.

Printing a part on a build platform covered with painter's tape.

The texture of the tape gives the filament good purchase so it sticks well. Some particular brands seem to work better than others, but most of the commonly available types found at hardware stores work pretty well.

Painter's tape seems to work best for PLA on unheated beds. It can be used with ABS and on heated beds, but there are usually better options. It's also not a particularly durable surface treatment, so you'll probably need to replace the tape often (maybe even after every print).

White Glue

Polyvinyl acetate (PVA)-based white glue (commonly referred to by the trade name *Elmer's Glue*) is also a popular way to get good adherence to the build platform. This is the same glue you probably used throughout your childhood for various craft projects, and it's easy to find at just about any office supply store.

To apply it to the bed, mix a small amount with water at a ratio of about 1:1. Next, use a small brush or paper towel to spread it around on the build platform; it can be applied on top of blue painter's tape or directly to the build platform. Ideally, you want a thin layer left behind as the water evaporates, leaving a sticky residue. The water doesn't have to be completely evaporated for you to start printing, but the build platform shouldn't be wet.

This glue helps the filament stick for obvious reasons—it's glue, after all! When used in conjunction with blue painter's tape, it can almost completely eliminate warping when printing PLA. However, ABS warps so much that you might not be able to completely overcome it. Also, be aware that you'll almost certainly need to replace the tape (if using) and reapply the glue before every print.

Polyimide Film

Polyimide film is used in a handful of particular industries and applications, including for superconductors and spacecraft. It's very strong (even in thin sheets) and flexible and can provide electrical insulation and thermal insulation. Polyimide sheets and tape are usually referred to by the trade name *Kapton* (developed by DuPont).

Polyimide film, like blue painter's tape, has a surface texture to which filament adheres very well. Additionally, it can withstand very high temperatures, and there is virtually no risk of damaging it with the heat generated by 3D printers. ABS also sticks very well to polyimide film, which makes it especially useful when combined with a heated bed.

A 12×12-inch sheet of polyimide film.

One of the most desirable properties of polyimide film, however, is its strength. It can be reused many times without damage, which is convenient of course. The only real downsides to polyimide film are its availability and cost. It's not usually available in local stores and has to be ordered online. And, even when ordered online, it can be fairly expensive.

PET Film

Polyethylene terephthalate (PET) film is used similarly to polyimide film, although its physical properties and traditional applications are quite different. PET film comes in sheets and tape like polyimide film and can be applied in the same way.

In fact, in the context of 3D printing, PET film is really very similar to polyimide film. However, it doesn't have the same ability to withstand high temperatures that polyimide film has. Polyimide film is often used to hold thermistors onto the hot end, for example, and this is an application PET film isn't well suited to. It just can't hold up to the heat from the hot end. But it works just fine for the build platform.

ABS Juice

Probably the most effective way to eliminate ABS warping is to use a sort of glue made from ABS and acetone, commonly called *ABS juice* or *ABS glue* in the 3D printing community. Because ABS dissolves in *acetone*, ABS juice is made by dissolving a large amount of ABS filament (new or from failed prints) in pure acetone, which can easily be found at your local hardware store.

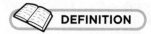

> **Acetone** is a common household and industrial solvent. Acetone breaks down styrene, which allows it to dissolve ABS.

When kept in a sealed container to keep the acetone from evaporating, this produces a viscous fluid. It can then be applied to the build platform, either directly onto glass or onto polyimide film or PET. This mixture is extremely sticky, and when a thin layer is applied, the acetone quickly evaporates away. After the acetone is evaporated, a film of ABS is left firmly stuck to the build platform.

Once you start printing, the molten ABS filament is deposited on top of this ABS film that is stuck to the build platform. ABS adheres well to itself, and so it sticks extremely well to the film (which remains stuck to the build platform). The result is a print that remains very firmly attached to the build platform.

Using ABS juice can almost completely stop ABS parts from warping, making it very useful. However, it can also be pretty messy to work with. It has to be made ahead of time, kept in a sealed acetone-safe container, and has to be applied to the build platform. And it works so well that it can be difficult to remove parts without damaging them.

Furthermore, the color of the ABS juice will coat the bottom of the printed part. So, for example, if you printed a white part with black ABS juice, the bottom would end up at least partially black. However, this can be counteracted by using a natural (uncolored) ABS filament when making the juice.

Hairspray

One surprising surface treatment that has recently gained popularity is hairspray. The same stuff that gave Farrah Fawcett's hair its famous flip can help your parts stick to the build platform. And, maybe even more surprisingly, it actually works very well!

As odd as it may sound, hairspray may just be perfect for 3D printing. It's cheap, it can be found in any grocery store or drug store, it's easy to apply, it works well, and it doesn't damage or add color to your parts. What more could you want?

The kinds of hairspray that work best are the ones that are marked with words like *extra* or *super*, or ideally *extra super hold*—the stronger the hairspray, the better. All you have to do is spray a few coats onto the build platform, wait for it to dry, and then start printing! However, it may help to remove the build platform from the 3D printer, if possible, in order to avoid getting hairspray on any of the moving parts or mechanisms.

Hairspray is essentially just an aerosol adhesive, so you're really just spraying a layer of glue onto your build platform. This works very well to keep printed parts stuck well to the platform, and when used on a heated bed, it can completely eliminate warping. Even large ABS parts can be successfully printed, which is pretty good for a beauty product.

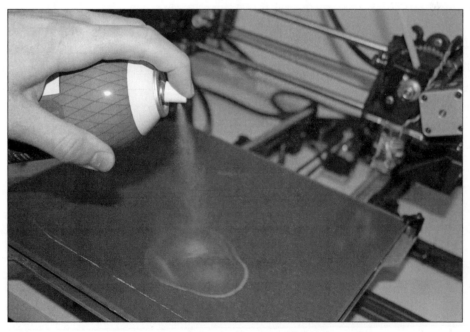

Hairspray being applied the print bed to help the part stick.

The Least You Need to Know

- Build platforms can be made from a variety of materials, as long as they're very flat.

- Heated beds aren't necessary for all materials but are required for some. However, they can improve print quality for virtually all materials.

- Heated build chambers can do a lot to reduce warping, but the technology is patented. Because of this, most consumer 3D printers do not have heated build chambers; however, many still use enclosures.

- There are many surface treatments that can be applied in order to improve the adhesion of the filament to the print bed, such as painter's tape, white glue, polyimide film, PET film, ABS juice, and hairspray.

CHAPTER
9

Control Components

Controlling a 3D printer is no easy task, because there is a lot of hardware to simultaneously coordinate. The movement of three axes has to be coordinated, the filament being fed by the cold end has to be controlled, the temperatures of the hot end and heated bed need to be monitored and adjusted, and fans have to be turned on and off. This all requires processing power, significant amounts of current being passed through the control board, and the simultaneous control of multiple outputs and inputs.

In this chapter, I explain what hardware is commonly used to control consumer 3D printers, some optional components you can use to help with the controls, and how they all work.

In addition to the required components that have to be controlled, there are also a couple of things that can be optionally used as well. An LCD screen and SD card reader, for example, can be used so it's not necessary to have a computer connected to the 3D printer for it to run. There is also one other type of component I haven't yet discussed, but which is very important: an end stop.

In This Chapter

- What end stops are and how they work
- What the control board does
- How LCD controllers and SD cards can help

End Stops

An end stop is a small and simple component with one very specific purpose: to tell the 3D printer when it has reached the end of each axis. This is a very important function, because it's the only way the printer can know where the zero point of each axis is. Without end stops, you would have to manually position the hot end each and every time you used the printer, which would be especially difficult for the Z axis (which has to be at an extremely precise height above the bed).

> **FASCINATING FACT**
>
> End stops are also used on CNC mills for exactly the same reason. A CNC mill, just like a 3D printer, needs to know where the zero point is for each of its axes in order to operate properly. The only major difference is that CNC mills don't have a constant zero point for the Z axis; instead, they need to find the top of the material being milled. For this reason, the Z height is either set manually or with a probe.

With such an important job to do, you might think that end stops would be expensive components. But in reality, they're actually generally very cheap parts. That's because they're simply switches—just very sensitive switches. The key traits of an end stop switch are sensitivity and repeatability. They need to be triggered by the faintest touch and at exactly the same distance every time. This is usually done in one of two ways: mechanically or optically.

Mechanical End Stops

Mechanical end stops are the simplest, cheapest, and probably most reliable type of end stop. They are, quite literally, just a physical switch that is triggered when each axis gets to its zero point. The switch consists of a small arm that sticks out from the body of the switch. When the extruder carriage, build platform, or Z assembly reaches the end of its axis, it pushes on the arm to temporarily trigger the switch.

The switch is either open or closed in its normal state, and the control board (which I'll discuss later in this chapter) monitors what state it's in. When contact is made on the arm, the state of the switch is reversed. So if it was open before, it's switched to closed. If it was closed before, it's switched to open. When the control board sees the state change, it knows that particular axis is at its zero point.

A switch wired to be normally open has a small amount of voltage applied to one contact of the switch. When the switch is triggered, the connection is closed and the circuit has voltage flowing through it. The control board sees the circuit has voltage flowing and knows that the switch has been flipped.

Switches that are wired to be normally closed work in the opposite way. When the switch isn't triggered, the voltage is always flowing through the circuit. But when the switch is triggered, it cuts off the flow. And, as expected, the control board sees that change and proceeds accordingly.

This is a common mechanical end stop switch, wired to be normally closed.

It's generally considered to be better to wire the end stop switch to be normally closed. This is because a 3D printer can be damaged if the switch isn't triggered and the axis keeps trying to move past its end point. With the switch normally closed, any damage to the circuit will result in the switch being triggered, so it acts as a sort of failsafe.

However, interference from the stepper motor wires can induce current in the end stop wires. So it's possible for the switch to be physically triggered but for the voltage to still reach the control board. If this happens, that axis could continue moving when it's not supposed to and could cause damage.

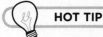

> **HOT TIP**
> Interference can be a major problem for end stop switches, either causing false triggers or failure to trigger. The interference is generally caused be stepper motor wires, which are often run right next to the end stop wires. To protect against this interference, the end stop wires should ideally be a shielded and twisted pair. A shielded and twisted pair is a type of wiring where pairs of wires are twisted together and then electrically shielded to prevent interference.

When the end stop is wired to be normally open, any induced current caused by interference would simply cause a false trigger, and no damage would occur. So both setups have their pros and cons, and one isn't necessarily objectively better than the other.

Optical End Stops

Optical end stops serve the same function as mechanical end stops. But instead of being triggered by physical contact, they are triggered by changes in light. This is handled by a component called a *photo interrupter*. No physical contact is necessary; instead, an optical end stop is triggered when the light in the photo interrupter is interrupted.

A photo interrupter works very much like the sensors used on automatic garage door openers. It consists of an emitter and a receiver (a phototransistor) that face each other. Infrared light is output by the emitter, and the receiver senses that infrared light. When it sees the light, the switch isn't triggered; however, when that beam of light is broken and the receiver no longer senses the infrared light, the switch becomes triggered.

This works almost exactly like your garage door sensors. When something is blocking your garage door, the infrared light beam is broken and your garage door opener knows not to close. The photo interrupters used in optical end stops do the same thing, just on a much smaller scale.

In a photo interrupter used for optical end stops, the emitter and receiver are mounted right next to each other. They're generally only a few millimeters apart, leaving just enough room for something to trigger the switch. This proximity makes them very sensitive, unlike your garage door sensors. This sensitivity is what makes them desirable for use in 3D printers.

As I've previously discussed, it's important for the 3D printer to know exactly where the end point of each axis is (especially the Z axis). Mechanical end stops do this job well, but because they're mechanical, there is always going to be small amount of difference (even if it's only a fraction of a millimeter). Optical end stops have no moving parts and are triggered as soon as the infrared light beam is broken. This makes them very precise, so the 3D printer knows exactly where the end point is every time.

Of course, optical end stops do have some drawbacks. The most obvious of these is the cost of the end stop itself. Mechanical end stops are just precise mechanical switches, making them very cheap. On the other hand, optical end stops use electrical components (namely the photo interrupter), which are more expensive. The cost difference isn't huge, but it is notable (though optical end stop prices have dropped significantly in recent years).

Optical end stops are also slightly more complicated to wire, as they require three wires (while mechanical end stops only require two). This is because they need constant power in addition to a signal wire. Mounting an optical end stop is also quite a bit more complicated than a mechanical end stop, because some sort of tab has to be used to interrupt the infrared light beam. This means the end stop has to be perfectly lined up with the tab on that axis in order to function, while a mechanical end stop can really be mounted in any way that lets it get pushed by something on that axis.

Other Kinds of End Stops

Because end stops are really just switches that get triggered when the axis reaches its end point, there are a lot of mechanisms that can be used as end stops. Anything that can act as a proximity switch could probably work, technically speaking. This means there are other types of end stops out there, even if they're not particularly popular.

One of these, for example, is a *Hall effect* sensor. A Hall effect sensor senses a nearby magnet and varies its output voltage based on the how close or powerful the magnet is. So a small magnet can be mounted on the moving part of a 3D printer axis, and when it gets close to the Hall effect sensor, the end stop is triggered.

DEFINITION

The **Hall effect,** discovered by Edwin Hall in 1879, is the tendency for voltage in a circuit to change when exposed to magnetism. It can be harnessed to create proximity sensors and is sometimes used for end stops and probes.

Beyond Hall effect sensors, people have experimented with a wide range of different types of devices to make end stops (everything from ultrasonic distance sensors to accelerometers). But so far, mechanical and optical end stops remain the most common types of end stops by a long shot. Their relatively low cost and reliability make them ideal for use in 3D printers, and there just hasn't been anything that works any better.

Control Boards

In order to monitor and control those end stops and all of the other components of a 3D printer, a control board is needed. The control board is a small computer that's essentially the brain of the 3D printer. It is responsible for both interpreting commands from the slicing software and signals from the 3D printer's sensors and using that information to control the operation of the printer.

Control boards come in a variety of different types and styles from many different manufacturers. There are open-source boards meant to work with various types of 3D printers, open-source boards meant to work with specific 3D printers, and proprietary boards that only work with the particular printer they were designed for. But they all tend to work in roughly the same way, with similar inputs and outputs.

Arduinos and Proprietary Control Boards

Every control board needs to be able to do basically the same things:

- Take the commands outputted by the slicing software and use them as instructions for operating the 3D printer
- Monitor and heat up the hot end (and probably a heated bed, too)
- Simultaneously control four or more stepper motors—one for each axis, and a fourth for the extruder
- Monitor at least three end stops

Doing all of this requires processing power, of course. But more importantly, it requires a lot of inputs and outputs. Every component on the 3D printer needs to have its own input or output (depending on its function), and they all have to be controlled or monitored at the same time. For this reason, Arduinos are a popular choice as the base of most open-source control boards.

Arduinos, by their very nature, are very well suited to this task. They're programmable, have onboard processing capabilities, and have native support for many inputs and outputs. When used for 3D printing, Arduinos are generally equipped with a *shield* that provides all of the connections for the steppers, hot end, end stops, and so on. Or, in some cases, the Arduino and connections are all integrated into a single circuit board for simplicity and ease of use.

DEFINITION

In the context of Arduinos, a **shield** is a circuit board designed to be attached to the Arduino board. Shields are generally used to expand the capabilities of Arduinos by adding either more connections or sensors. They are usually designed for a specific application (like GPS tracking or to interface with another device), although plenty of general-use shields exist.

A RAMPS 1.4 control board, which is a shield for an Arduino Mega.

Proprietary control boards that are closed source and aren't Arduino based still work in the same way but are just developed by the manufacturer specifically for use in a 3D printer. They still have to have all of the same capabilities for processing and input and output; the only difference is they aren't built on an Arduino. It's analogous to internet browsers: whether you're using an open-source browser like Firefox or a closed-source browser like Internet Explorer, they display web pages, and you can use them to surf the internet.

Changing or Upgrading Control Boards

In many cases, the control boards can be interchangeable. For example, if you decided you wanted to add dual extruders to your printer but the control board didn't have the proper support, you could potentially remove the original control board and replace it with one that supports dual extruders. This is possible because the other components of the 3D printer all tend to be the same, regardless of who makes it, and they work in the same way.

However, complications can be encountered when it comes to firmware and software. Firmware is made specifically for the control board; it can't be transferred between different kinds of control boards. And in some cases, only specific slicing and control software can be used with a particular control board. This isn't the case with open-source control boards, which virtually always work with the common software programs, but it can be the case with some proprietary software and control boards.

So when should you worry about the control board, if ever? There are generally two instances where the control board used by the 3D printer might make a difference to the user: when it comes to expandability, and if the software you can use is constrained.

WATCH OUT!

While it's usually possible to replace the control board with a different type, you shouldn't count on it. Changing the control board can be complicated by the enclosure and connectors used. It's rarely a plug-and-play type of change, even when it is possible.

The expandability of the control board can be a pretty big deal if you ever want to upgrade or modify your 3D printer. If you want to add dual extruders, a heated bed, an LCD controller, or anything else, the control board has to be capable of supporting it. Some control boards simply don't have the connections necessary to add these things. Or, even if the physical connection is present, there may not be firmware available that supports the control of those parts.

Software compatibility can also potentially be a fairly major concern. Some software will only work with specific control boards, and vice versa. This is usually only the case with 3D printers from manufacturers who want a complete 3D printing ecosystem under their brand. They develop a control board specifically for the printers they manufacturer, and then have slicing and control software which just works with those control boards.

In such a situation, it may be difficult or impossible to use any other software to slice 3D models or to control the 3D printer. It also may not be possible to use the software if you don't have that particular 3D printer. The latter problem may not big a very big deal, because there are plenty of software options that are free and work well. But the former problem could be pretty serious. If you don't like the software released by the manufacturer, you'd still have to use it anyway. If the manufacturer went out of business or stopped supporting the software, you may be stuck using outdated software.

If either of these things sounds like something that might matter to you, it's a good idea to do a little research ahead of time about what you're getting. Does the control board allow you to add upgrades later? Are you free to use whatever software you like, or are you stuck with a particular program made by the manufacturer? Depending on your needs and expectations, those could be major factors in your decision on which 3D printer to purchase.

SD Card Support

One nifty add-on that many control boards can take advantage of is secure digital (SD) card support. Normally, you have to have a computer connected to a 3D printer while it runs. This is so the host control software can send the instructions to the printer, which tells it what to do to create the part. However, those instructions are created ahead of time by the slicing software. All the host software does is control the setup of the printer (like homing the axes and heating the hot end) and then send the instructions via USB to the 3D printer. At this point, the computer isn't really doing any processing, other than what it takes to run the computer itself and the software. All it's doing is storing the instructions on the hard drive and then sending each command to the printer.

Because no significant processing is being done, the computer really isn't necessary after the slicing has been done. Instead, the instructions can be saved to an SD card and the control board can read the instructions directly without having to use a computer as an intermediary. Some control boards have this functionality built in, while others can have it added as an upgrade later. (Some control boards don't have this ability at all, but luckily, they're in the minority.)

Using an SD Card

If the control board does come with SD card support, or it has been added, you can forgo the use of a computer entirely during printing. You still have to use slicing software on the computer to convert your 3D model into a series of instructions that can be read by the control board. But once those instructions have been created, you don't have to use a computer at all; you can simply save the instructions from the slicing software on the SD card and then insert it into a slot on the control board (or a separate board, if SD card support is added).

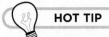

HOT TIP

If your control board does support SD cards, make sure you do a little research on what kind it supports. SD cards come in different physical sizes, and you need the right size for your SD card reader. Additionally, they come in different storage capacities, and some might not be compatible with the control board if they are too large.

When the 3D printer is turned on with an SD card inserted, the control board will automatically look for the instructions created by the slicing software. If it finds the instructions, it will proceed with running them. Generally, the instructions include information on what temperature the hot end and heated bed are supposed to be and any homing instructions. So the 3D printer will start with those things, and once it's heated up to the proper temperature and homing has been done, it can start printing.

From then on, it's just like if you were printing from your computer, because it's reading the exact same instructions from the SD card that the computer would have been feeding to it over USB.

Benefits of SD Cards

SD card support has a number of benefits that might prove very useful. There is the obvious benefit of not having to leave your computer attached to your 3D printer whenever a print is running. That way, you can use your computer for other things while your 3D printing is printing. You don't even have to have the computer in the same room (or even building) as the 3D printer, because the printer only needs the SD card to run.

It also makes printing more reliable, because the computer is no longer part of the equation. Anyone who has ever had a long print ruined because Windows Update decided to restart the computer knows how valuable that can be. The same benefit applies for those of us who have dogs that like to pull out USB cables.

For people who are eco-conscious (or budget conscious), having SD card support means you don't have to waste electricity running your computer. If you're frequently doing long prints, this can actually be pretty significant. Leaving your computer running for two days while your 3D printer makes some big object can use quite a lot of power.

LCD Controllers

As beneficial as SD card support is, it does present a couple of challenges. Because it can only read instructions that have already been created by the slicing software, that means you still can't manually control the printer without a computer. There is also no way to choose from multiple instruction files, which means you can only store one file on the SD card at a time to print.

Luckily, there is a solution for that: an LCD controller. An LCD controller combines an LCD panel for displaying information, as well as some sort of control mechanism (or it could be a touchscreen LCD). Many LCD controllers also integrate an SD card reader or can be used with a separate SD card reader. In either case, the result is the same: you can manually control the 3D printer and select specific files from the SD card to print.

An LCD controller with SD card support made by RepRapDiscount.

When a 3D printer is equipped with an LCD controller, you can manually move the steppers (including the extruder) and heat up the hot end and heated bed without using a computer at all. When an SD card is inserted, you can manually select which files you want to print as well. So you can load up the SD card with many files and choose which one you'd like to print.

This makes an LCD controller very useful in situations where you don't have a computer nearby. For instance, at trade shows and conventions, you could bring the 3D printer with an SD card full of example prints. You could then have your 3D printer churning out parts at a booth without even having to bring a computer.

It's also useful in situations where a 3D printer is shared between multiple people, like at a school. Students can use their own computers to slice models and then load the files onto an SD card. The students can then load their files from the SD card without having to have a computer dedicated to controlling the printer.

FASCINATING FACT

What if everything could be handled by the 3D printer without having to touch a separate computer at all? This is territory that is just starting to be explored. A small touchscreen computer can be mounted on the 3D printer, allowing you to download 3D models, slice them, and print them without ever touching your computer.

This is still a pretty new technology, but many 3D printer manufacturers are working on integrating it into higher-end 3D printers. It isn't cheap, of course, but with the popularity of tablets, it's not as expensive as you might think. And the usefulness of such a setup is so high that it's likely to become commonplace in the next few years.

The Least You Need to Know

- End stops are necessary for telling the 3D printer when each axis has reached its end point.
- Mechanical and optical end stops are by far the most common, but any type of proximity sensor could technically work.
- The control board is responsible for interpreting the instructions created by the slicing software and for using those instructions to control all of the components of the 3D printer.
- Some additional devices, like LCD controllers and SD card support, can be added to control boards to improve their functionality.

CHAPTER 10

Choosing a 3D Printer

Now that you know the different parts that make up a 3D printer, it's time to think about what type of 3D printer is right for you. Deciding on what 3D printer to purchase can be both overwhelming and exciting. The increasingly abundant number of manufacturers and models on the market are sure to make anyone dizzy. Do you need a 1.75mm or 3mm hot end? Is one company better than another? Should you care about whether or not it's open source?

There are a plethora of options and factors to think about, but if you can make a few early decisions, you'll gain a lot of focus in your search. Most people already have an idea of what they want and expect from their printer, so it's just a matter of consciously deciding what's important to you.

In this chapter, I take you through some things you should consider when purchasing a 3D printer. Once you've narrowed your list down to a few models, the choice will become a simple matter of picking which has the best reviews and price.

In This Chapter

- Open vs. closed source
- Whether getting an assembled printer, printer kit, or DIY printer is right for you
- What are your printing needs?
- Designs beyond Cartesian printers

Open or Closed?

Do you care if the printer is open source? For many people, this is the most important factor in their purchase. For others, it's largely irrelevant. The consumer 3D printer market is divided roughly 50/50 between open-source and closed-source printers, so choosing which kind you want initially can cut down your options by half.

What Does It Mean to Be Open Source?

Let me first explain what I mean by "open source." If you recall from Chapter 2, the basic open-source concept is simple: *intellectual property* is shared freely so others can modify it, use it, and share it themselves. This is in direct opposition to the practice of controlling intellectual property via patents and copyrights (otherwise known as closed source). Open-source data can still be patented and copyrighted, but to be considered open source, it has to be licensed to allow sharing.

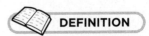

> **Intellectual property** is any idea that is legally protected. Providing legal protection for intangible things like ideas is a complicated matter, and the laws vary from country to country. But in the United States, intellectual property like inventions, music, copywriting, patents, and so on are legally protected property.

In the early days of the internet, many people believed that patents and copyrights of technology, especially software, were counterproductive and inhibited progress. Both individuals and corporations pushed the idea you could share your ideas while making a profit and fueling innovation.

In 1985, the Free Software Foundation (FSF) was created to support and promote this movement. At the time, the primary goal was to ensure people's freedom to use their hardware as they pleased without being restricted by proprietary software. Over time, this ideology evolved to encompass a broader idea of open software. Then, in 1998, the Open Software Initiative (OSI) was created with very similar goals in mind. Together, these organizations (and others) have helped to create the many free and open software options we enjoy today.

But it doesn't stop with just the software. A more recent trend is the open hardware movement. The ideals are similar, but instead of developers making their source code public, engineers are making their designs public. In the same way you can download Linux source code under the GNU License created by FSF, you can download blueprints and design files from open hardware companies. The entire RepRap project is based on that idea and has been completely open source from the very beginning. In fact, many of the 3D printers on the market today are RepRap designs or derivatives.

Additionally, other manufacturers have developed their own designs independently of the RepRap project but also have made them open source. For example, Aleph Objects makes the only 3D printer to carry a hardware certification from the FSF. That printer, the Lulzbot, is the printer featured in the photos throughout this book. The kind of information most companies keep under lock and key, such as design files and documentation, is made freely available by Aleph Objects.

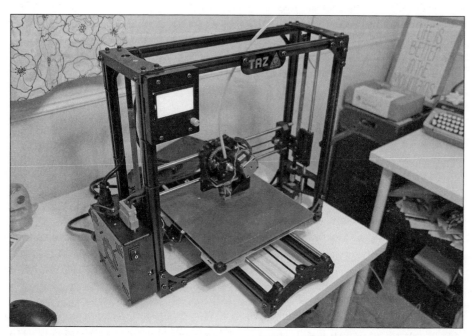

The Lulzbot TAZ 4 is the only 3D printer available that is certified by the FSF.

Many other manufacturers stay true to the open-source ideology as well. Like Aleph Objects, they make their design and engineering documents available to the public. The key is they share their designs, make their printers compatible with open-source software, and allow customers to use their printers however they wish.

Why It Might Matter to You

You now know what it means to be open source, but why should it matter to you? What difference will it make, practically speaking?

The biggest argument for the purchase of an open-source printer is freedom. You'll have the freedom to use whatever software you please and to modify your printer however you like. You won't be restricted in what you can and can't print, where your files come from, or how you use them. If you decide you don't like the software recommended by the manufacturer, you can just switch to another program. You won't be required to use proprietary filament cartridges or be restricted to approved accessories.

 **HOT TIP**

Many 3D printers with closed-source hardware are still capable of using open-source software for slicing and control. Open-source software may even be recommended by the manufacturer. When you're looking at printers, be sure to check which software the manufacturer recommends and whether you're free to use other software.

With all of that said, there are many great 3D printers on the market that are not open source. You may decide the printer for you is closed source. There is absolutely nothing wrong with going that direction. After all, the vast majority of things we buy on a day-to-day basis aren't open source. Most completely closed-source manufacturers strive to make their systems seamless and easy to use, and you may find that you prefer one of those systems. But be sure to consider the decision carefully, since closed source is more restricted.

Assembled, Kit, or DIY?

The next factor to consider is how you want to receive your printer from the manufacturer, or if you want to skip the manufacturer altogether and build your own 3D printer. This choice shouldn't be taken lightly, so I'm going to explain the advantages and disadvantages of each option.

Assembled 3D Printers

If you are unsure what you want to do, this is absolutely the safest choice. Purchasing your printer already assembled from the factory is the easiest and most reliable way to start printing. You won't have to worry about assembly; you can just unbox it, do a few setup steps, and be printing your first model in an hour or two. And most assembled printers will come with a warranty of some kind in case anything goes wrong, which may not be the case with the other options. It's also likely that you'll be able to receive some technical support if you run into any problems.

Cost is the biggest downside here. Someone had to be paid to assemble the printer, and that results in a higher price. Shipping costs may be higher as well, as the box will usually be bigger than that of an unassembled printer. And, although it's assembled, you'll probably still need to do a little bit of adjustment to get it actually printing.

3D Printer Kits

Kits are a great way to save some money and an opportunity to learn about your printer as you're building. The knowledge and experience you gain building the printer can help tremendously in the future with troubleshooting and modifications.

Generally, a kit will include all of the parts and hardware you need to build your own printer. It should come with the frame pieces, fasteners, electronics, motors, and anything else you'll need to build a working 3D printer. Most kits on the market today are based on RepRap designs, but there are some that were independently designed. Like with an assembled printer, your decision should be based on your needs, but you'll also need to take into account build time and difficulty.

What it takes to build a kit can vary quite a lot. Some kits come partially assembled and just require that you fit the pieces together. Other kits will require that you assemble each and every part on your own, so you may need special skills (like the ability to solder). You may also have the option of ordering a partial kit that includes everything except 3D-printed parts, which can be a good way to save some money if you know someone who can print those parts for you.

The downsides to kits are that they can be very time consuming to assemble, they require tools, and assembly can be difficult. Therefore, make sure you're confident in your ability to assemble the printer before you purchase it.

Building a DIY 3D Printer

Another option is to build your own do-it-yourself (DIY) printer using designs published by individuals or open-source companies. Most of these designs are based on the RepRap project, and many are tried and true. Building your own printer can save you a significant amount of money and gives you the opportunity to modify the design to build the exact printer you want. (See Appendix B for some helpful links on how to do this.)

My DIY 3D printer, based on the open-source TAZ design.

The downside, of course, is that it's completely up to you to source all of the parts and put them together. This is a very difficult task, and I wouldn't recommend it for your first printer. But for a second printer, or for the adventurous, it can be very rewarding and a great learning experience.

Determining Your Needs

At this point, your choices should already be considerably narrowed, and hopefully a few manufacturers and models are starting to stand out. When looking at these remaining models, consider what your needs are and what you need your printer to do for you. What kind of models will you be printing? What will their use be? What kind of filament will you use? What should everything cost? The following are a few specific things you'll want to pay attention to.

HOT TIP

3D printer manufacturers will sometimes publish specifications in inches, millimeters, or both, so you may have to do some unit conversions when comparing printers. However, metric units (millimeters specifically) are generally the standard when exporting and importing STL files and when configuring parameters in software and firmware. Be sure to check what units are used and required when working with different 3D printers.

Print Volume

Print volume is the specification that tells you how large a model (or models) you can print. In most cases, this will be rectangular, and is calculated by multiplying the print width, length, and height. However, for some printers (particularly Delta-style printers), the print volume is cylindrical and calculated by multiplying pi by radius squared and height ($\pi r^2 h$). If you do choose a Delta printer, be sure the circular base is suitable for your needs.

While looking at the print volume is certainly a useful way of comparing printers, it doesn't quite tell the whole story. Two printers with the same print volume could have vastly different dimensions that might affect your decision. For example, Printer A has print dimensions of 8×8×8 inches, while Printer B has print dimensions of 16×16×2 inches. Both printers will have a print volume of 512 cubic inches, but how useful would a printer be that could only print a model 2 inches tall, as in the case of Printer B?

For that reason, it's a good idea to look at the actual dimensions. What is the print size in the X direction? How about Y and Z? Determine the largest size you'll be likely to print, and make sure the printers you're considering can handle it.

You may be tempted to just buy the printer with the largest print volume, just in case. But keep in mind that both cost and print time increase with print volume. Therefore, there is little sense in paying a lot of money for a big print volume you won't actually use. Plus, print time increases exponentially with size (you can thank the *square-cube law* for that).

> The **square-cube law** is a scientific principle that describes the mathematical relationship between size, surface area, and volume. Its relevance to 3D printing is the way that volume (and therefore print time) increases exponentially with size. For example, a cube 10mm to a side has a volume of 1,000mm³. Doubling the dimensions of the cube to 20mm to a side results in a volume of 8,000mm³. Because the larger cube is 8 times the volume, it will take 8 times longer to print than the smaller cube.

Print Resolution

Print resolution is essentially a measure of the potential quality of your printed models. Smaller is better here, similar to how smaller pixels result in better image quality on your computer monitor. A smaller resolution means you can print finer details, and the overall surface finish of the model will be smoother. But determining the print resolution that a printer is capable of can actually be a tricky task. There is currently no practical way of specifying the actual effective print resolution of a 3D printer in the real world.

Theoretically speaking, print resolution should be the result of two factors: nozzle size and movement precision. The nozzle size determines how fine the filament thread being ejected from the hot end is. Because nozzles are replaceable, this isn't something you should be too concerned with—you can always change the nozzle to one with a smaller opening. However, movement precision can't be altered without significant modification to the printer's hardware, making it an important consideration.

This can be calculated with the "steps per mm" information, which should be provided by the manufacturer for use when configuring your slicer and firmware. For instance, if your X axis is 200 steps per mm, your movement precision on that axis is 0.005mm (1mm divided by 200). But no matter how precise the movement is, the nozzle size will restrict how detailed your prints can be. If the nozzle is too big, it won't be able to print very fine detail. It's analogous to trying to write small text with a marker—the marker is just too big to write fine details.

There are other concerns that factor into the resolution of a 3D printer though. Just something that seems relatively minor, like flex in the printer's frame, can reduce the effective resolution of the printer. This means that the resolution specified by the manufacturer might only be achievable at slow speeds under the best circumstances.

Filament

There are many filament materials on the market, and the choices are rapidly increasing. Different 3D printers are capable of utilizing different filament materials, and it's worth paying attention to this when choosing a printer.

If you remember from Chapter 7, the two most common filament materials in use today are PLA and ABS. PLA is widely considered to yield higher-quality prints, while ABS parts are generally stronger and more durable. ABS is also easier to work with postprint (for sanding, drilling holes, and so on). Virtually all FFF 3D printers are able to print with PLA, while ABS requires a heated bed to print successfully. If you purchase a printer with a heated bed, you'll be able to print the two most popular materials. However, 3D printers that don't come with a heated bed included will often have that as an accessory specifically to allow you to print ABS.

Other more exotic materials may require specialized hardware. For example, if you're interested in printing *flexible filament*, a printer with a Bowden extruder will not work. The filament simply bends too much in the feed tube and won't be able to feed reliably. So you should only consider direct feed extruder types if you want to print flexible filament.

DEFINITION

Flexible filament is an umbrella term for any filament material that is flexible, rubbery, and squishy. This can be used as an alternative to the hard and rigid plastic that is more commonly used for 3D printing. The exact composition differs depending on the manufacturer, but the resulting material is fairly similar with all of them.

Other materials, like polycarbonate, require special hardware to print. In Chapter 18, I'll go into a lot more depth about what other materials are available and what kind of hardware you'll need to use them. It's worth skipping ahead to that chapter if you're curious about some of the more unique materials and want to know what you'll need to print with them.

Besides filament material, you also need to decide on what filament size you'd like to be able to use. Currently, 3D printer filament comes in two sizes: 1.75mm and 3mm. Originally, most filament came in 3mm. As nozzle sizes decreased, however, it took more force to push 3mm filament through the nozzle. For that reason, 1.75mm has become the more popular choice, with a wide range of manufacturers and color options. However, flexible filaments generally feed more reliably in 3mm because the thickness increases the stiffness, making it less likely to bend and jam. Some filament manufacturers only sell their flexible filaments in 3mm due to that.

With all of that said, both 1.75mm and 3mm 3D printers will have a very wide selection of filament to choose from. Don't put too much weight on this factor; rather, first find a 3D printer you like. If that printer has an option of 1.75mm or 3mm, decide on a size based on your needs.

Prices

Until just a few years ago, the high cost of 3D printers meant that usually only corporations could afford to purchase them for prototyping purposes. Luckily for both hobbyists and small companies, 3D printer prices have recently dropped dramatically. In a very short amount of time, the least expensive printers have gone from being tens of thousands of dollars to just a few hundred. Even better, a current midrange consumer printer is capable of rivaling the quality and print volume of professional 3D printers from only a few years ago.

So what can you expect to pay for a consumer 3D printer for personal use? Obviously, price will vary depending on size and features, but proven and well-reviewed printers range from about $400 to $3,000. At the bottom end, you have very basic printers with small print volumes that are often constructed out of inexpensive materials like wood. As prices rise, print volume generally increases, more features are added, and frames are made of higher-quality materials. At the high end of the consumer market, you have 3D printers that would even be suitable for an engineering or prototyping environment.

Simply put, this is an exciting time to purchase a 3D printer! Prices are low enough that it's realistic for an average person to purchase one for home use. The market is very healthy, and there are options out there to suit a variety of needs.

HOT TIP

When thinking about cost, remember to keep room in your budget for tools, accessories, and filament as well. Filament prices range from about $25 to $50 per kilogram for ABS and PLA, and a kilogram should be enough for you to get started with. It's also a good idea to have a basic set of hand tools available—such as wrenches, pliers, and screwdrivers—along with various glues and tapes.

Printers with Unusual Designs

For the purposes of simplifying this guide, I've been talking primarily about Cartesian 3D printers that use FFF technology for printing. Cartesian printers are set up with a rectangular build area with movement in three axes, and are by far the most common type of printer on the market (see Chapter 5).

However, there are other kinds of printers out there which you might come across. One type that has gained a lot of popularity recently is the Delta-style 3D printer. These use a unique method of movement where the build platform is stationary and the hot end is moved on three arms attached to vertical rails. Delta printers are advantageous because they're capable of printing at higher speeds. However, being a newer design, there is less information and support available. They also have a circular build platform as a result of their unique geometry, which might not suit your needs.

Chapter 10: Choosing a 3D Printer 123

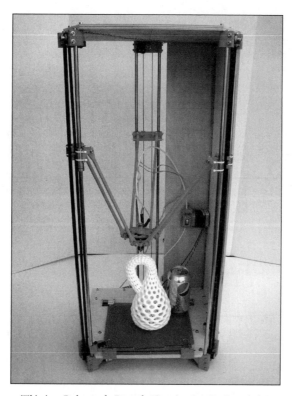

This is a Delta-style Rostock 3D printer, a RepRap design.

Another design that has recently started to enter the consumer market is digital light processing (DLP). Traditionally, these types of 3D printers were expensive and the technology was only used in professional printers. But prices have dropped low enough that they could be worth considering. DLP printers work by shining light onto a vat of UV-curable resin. As the light hits the resin, it cures and hardens, eventually resulting in a plastic part.

DLP 3D printers are capable of producing high-quality parts with very fine detail, but they also have significant drawbacks that make them unattractive options outside of specialized uses. The resin used is much more expensive (per kilogram) than traditional filament, and the selection is much smaller. Plus, color and material options are only a small fraction of what is available to FFF printers. They're also unable to print hollow parts, as the uncured resin is left inside the part.

In addition to Delta and DLP 3D printers, there are a number of other more exotic designs out there. However, they are all fairly experimental at this point and probably wouldn't be a good choice for your first 3D printer. In fact, it's probably best to stick to a Cartesian printer unless you have a specific need for one of the other types. The popularity of Cartesian printers ensures that you'll have access to a wealth of information and support, as well as a wide range of choices.

The Least You Need to Know

- Make sure you consider your present and future needs when determining what printer to buy.
- Choosing an open-source 3D printer provides benefits for support, flexibility, and modifications.
- Prices have dropped dramatically in recent years, but 3D printers are still a hefty investment. Kits offer savings and an opportunity to learn the ins and outs of your printer.
- If you're new to 3D printers, it's best to stick with the popular Cartesian printer. That way, you'll have access to plenty of support and information.

PART

3

Setting Up and Printing

Now that you're familiar with how 3D printers work, it's time to get into their actual operation. In this part, I go over how to set up your 3D printer's hardware when you receive it. I also teach you how to configure your software and connect to your printer.

You also learn important things like how to level the bed and how to adjust the height of the Z axis. I then finish this part with some basic printing projects to get you started.

CHAPTER 11

Software Setup and Printer Control

Before you can actually print anything, you have to set up the software to slice models and control the 3D printer. There are many options for these programs, and it's possible that your 3D printer may be restricted to one particular software package. However, the settings and setup process tend to be fairly similar between all of them.

In this chapter, you learn about your printer's firmware, as well as how to choose host and slicing software. I then discuss how you use this software to connect to and control your printer.

In This Chapter

- The purpose of firmware and host software
- How slicing software works
- Common host control commands and useful G-code

Firmware Explained

Firmware is basically a program that is stored on a device. I know that sounds terribly vague, but that's because the specifics vary quite a bit depending on the device. For example, on your computer, the operating system is not firmware, but the *Basic Input/Output System (BIOS)* is. On a cell phone, however, the entire cell interface may run on firmware, or the firmware could just be used to load an operating system like your computer.

The key characteristics of firmware are the following:

- It's always the first thing to run when the device is turned on.
- It's not stored in mass storage. Instead, it's stored on an onboard chip that's always present, since devices can't function without it.
- It's almost always set up for only one specific device.

> **DEFINITION**
>
> The **Basic Input/Output System (BIOS)** is the firmware interface used on most computers. It's the first thing to load as soon as your computer starts up and controls how the operating system (Windows, for example) is booted.

Firmware is different from software, though in a very general way they appear similar. Think of it as being as series of layers built on top of each other: the firmware loads first and directly controls the hardware, and then the software loads and only controls the hardware indirectly via the firmware. Because the firmware controls the hardware directly, it has to be written for the specific hardware being used. This allows the software to be much more generic (and to work with a wide range of devices), because it's not directly interfacing with the hardware.

In 3D printing, the firmware is stored on the control board and is used for the complete operation of the 3D printer. No software is needed on top of the firmware, and all of the functions of the 3D printer are controlled by the firmware running on the control board. Despite the complexity of 3D printers, the software needed to control them is still small enough to be loaded as firmware onto the read-only memory (ROM) of the control board.

Because firmware has to be small enough to fit in ROM, it's generally written to be as small as possible. The small size combined with the base-level operation of firmware usually means it can only work with the specific hardware it was designed for. If the hardware is changed, the firmware has to be rewritten for that hardware.

In the case of 3D printers, there are many types of printers, with many types of control boards. Luckily, they're designed similarly enough that the same basic firmware can be used for a lot of them. The firmware just has to be properly configured for the control board it's loaded on. That configuration can be performed by the printer manufacturer or even by the user. However, other control boards—especially ones that aren't open source—can only be loaded with firmware released by the manufacturer. If that's the case, the user has to rely on the manufacturer to provide adequate firmware.

FASCINATING FACT

There are only a few basic firmware systems that are used for the majority of 3D printers (especially open-source 3D printers). This is possible because the hardware used for 3D printers is so similar for most models. As long as the control board is compatible, the firmware just has to be configured for the specific hardware being used.

Still, 3D printer control boards almost always have firmware that is somewhat configurable even after it has been loaded. This is so the user can adjust the settings for things like the size of the printer, how many steps it takes to move each axis 1mm, and so on. This information is stored in the electrically erasable programmable read-only memory (EEPROM) and can be changed and overwritten.

If new firmware is flashed (essentially installed) on to the control board, the data stored in the EEPROM is still kept, so the settings remain. This is good for when you're just updating firmware, but keep in mind that if the EEPROM settings are incorrect, they will still persist. If you notice some problem doesn't go away even after flashing new firmware, you'll probably want to make sure it isn't some EEPROM setting that is causing the problem.

Unless you have a problem with it, though, there is usually no reason why you need to do anything with the firmware unless you want to. Occasionally, manufacturers will release new firmware versions that either fix a bug or add a feature, but for the most part, the firmware isn't something you need to worry about as a user.

Choosing Host Software

Unless you have an LCD controller (see Chapter 9), you need host control software to manually control the printer and run prints. The host software is what controls communication between your computer and your 3D printer.

The host software connects to the 3D printer's control board via USB. When you start a print, the host software takes the commands generated by the slicing software and sends them over the USB connection to the 3D printer. The number of individual commands can easily number in the thousands, and each command is sent one at a time. Once the command has been performed, the 3D printer's control board returns a confirmation to the host software, which then sends the next command. These steps are repeated over and over again until the print is finished.

Some 3D printer manufacturers require specific host software to control their printers. In most cases, however, you can choose from a variety of host software, including Repetier, Cura, ReplicatorG, and Pronterface. All of these perform the same general functions: manual control of the 3D printer, sending commands to run prints, and usually integration of slicing software. You can also modify EEPROM settings via the host software.

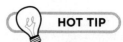

HOT TIP

Because each command is sent one at a time, the USB connection has to remain established throughout the entire printing process. If the USB cable is unplugged, the host software is closed, or the computer is turned off, the printing will stop and the part will be ruined. For this reason, you should make sure your computer is running throughout the entire print. More than one part has been ruined by a Windows Update restarting the computer halfway through the print.

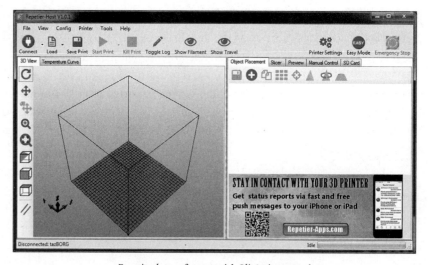

Repetier host software, with Slic3r integrated.

Choosing Slicing Software

Even if your 3D printer has an LCD controller and host software isn't needed, you'll still have to slice your 3D models for printing. That's what slicing software is for: to convert your 3D models into a series of *G-code* commands. This is a complex task that requires a lot of processing power, so for now it's necessary to do the slicing on a computer.

> **G-code** is the programming language that 3D printers and other computer-controlled machine tools can use for instructions. The G-code is what is used to give most 3D printers the commands they follow to produce parts. I'll go into more detail on G-code later in this chapter.

Unlike host software, the slicing software you use can actually make a pretty big difference in how your prints turn out. This is because the slicing software is what determines how the 3D printer actually moves to create the part.

Slicers use algorithms to break down a 3D model into a series of steps for the 3D printer to follow. Those algorithms are all programmed differently, so one slicer will produce different commands than another slicer would. This can result in very different prints, depending on which slicer you choose to use. Some may produce higher-quality prints, some may print faster, and some may handle special tasks (like generating supports) better.

Because the slicing software can make such a big difference in how parts are actually printed, it can certainly be worthwhile to try multiple slicers. For example, some slicers might handle the curved surfaces of artistic models better, while others might be better suited to engineering models that have flat surfaces and require a lot of supports. You might even find that some slicers work better with your particular 3D printer than others. Experimenting with some of the popular slicers can help you find the one that suits your priorities, the 3D printer you're using, and the kinds of models you'll be printing.

You can find slicing software integrated with host software in the same package. Repetier and Cura are two examples of host software that include slicing software with them. However, slicing software that's separate from host software is also available, such as KISSlicer, Simplify3D, and many others.

Connecting to Your Printer

Once you've chosen which host software and slicing software to use (or at least to try), you need to actually connect the host software to your 3D printer. How you do this depends on your printer, your control board, and the host software you're using.

If you've purchased a printer that requires the use of proprietary software, I can't give any real specifics on how to connect it. The good news is that 3D printer manufacturers usually choose to use proprietary software in order to make the user experience friendlier, so connecting your printer should be a simple process. The manufacturer should also provide you with instructions on how to do so.

However, if you purchased a printer with an open-source control board, there are some steps you can follow to get your 3D printer connected:

Get the proper drivers installed for your control board. The drivers give your computer the information it needs to establish a USB connection with your 3D printer. These are drivers just like you'd have to install when you connect any other new device to your computer, like a camera or an inkjet printer.

Which drivers you need to install not only depend on your operating system, but also control board you're using. Arduino-based control boards, for example, need Arduino drivers that can be downloaded from the Arduino website (arduino.cc/en/Main/Software). Check with the manufacturer of the 3D printer or control board for the specific drivers needed for your printer.

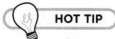 **HOT TIP**

If your control board is Arduino-based, you'll probably want to go ahead and install the complete Arduino software package. This will allow you to modify and update the firmware in the future if you choose to, along with providing the drivers needed to connect the 3D printer to the computer.

Connect a USB cable from the 3D printer's control board to your computer. Your computer should give some indication that it sees the USB connection but shouldn't try to install anything. If it tries to install drivers, the drivers might not have been installed properly before; go back to the first step to get your drivers in order and then try connecting again.

Open your host software and choose the COM port. Somewhere on the interface, you should see a way to change the printer setting and a button to connect the printer. In the settings, choose which COM port your 3D printer is on. This should correspond to which USB port the cable is plugged into, but there is generally no physical indication of the numbering used for the USB ports.

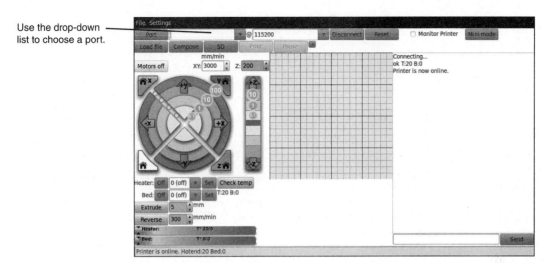

Use the drop-down list to choose a port.

Connecting to the correct COM port using Pronterface.

You can check your device manager to determine which COM port is being used by the 3D printer, but sorting through the device manager can be difficult, as the devices aren't always named as you might expect. Instead, it might be easier to try each of the COM ports if you don't have a lot of devices connected via USB. If you don't see the COM port which corresponds to the USB connection with your 3D printer, refresh the list of available COM ports.

Once you've found the correct COM port, it should only take a few seconds for your printer and host software to connect.

Controlling Your Printer

Once the connection between the host software and the 3D printer has been made, you can manually control the printer. Your host software should have an interface for doing this, usually called something like *manual control*. This is for manually sending G-code commands to the 3D printer's control board, which then follows the command. (I'll go over G-code more later in this chapter.)

Most host software have a few basic manual control functions, the most common of which are the following:

Jog axis: This command is used to move an axis by a set amount in a particular direction. The movement distance is usually set in increments—like .1mm, 1mm, 10mm, 50mm, and 100mm—in both the positive and negative directions.

 **WATCH OUT!**

If you don't have end stops on your 3D printer, do not use the homing functions. Without end stops to tell the 3D printer when the axis is at its end point, they will keep trying to move even when they can't physically move any further. This could potentially damage your printer.

Home axis: Homing moves the extruder, print bed, or Z axis assembly to its end point. When this function is used, the printer moves the selected axis until its end stop is triggered. Once the end stop has been triggered, the printer sets the current position for that axis to zero.

Home all: This function simply homes all of the axes with one command. Generally, it homes each axis one at a time, not all simultaneously.

Extrude: This command is for extruding a set amount of filament. You'll generally only need to use this command when loading new filament or when calibrating your extruder.

Retract: This is the opposite of the extrude command. It runs the extruder stepper motor in reverse in order to pull the filament out of the hot end. Often, the extrude command and retract command are combined, as retraction is really just negative extrusion.

Stepper motor power: Normally, once you've started manually controlling the 3D printer, the stepper motors are given power. Even when they're not moving, the stepper motors use power to hold their positions. If you want to physically move anything connected to a stepper motor, you have to turn off the power to the stepper motors, which is what this command is used for.

Hot end power: This provides the obviously necessary function of turning the hot end on and off and setting the temperature of the hot end. If you're going to be manually extruding filament, you have to first heat up the hot end to the proper temperature.

Heated bed power: As you might expect, this turns the heated bed on and off and allows you to set its temperature. There really aren't many reasons why you'd need to manually turn on the heated bed, but it can be useful for preheating the bed before you start a print.

Fan power: I'm not sure why you'd want to manually turn on the fan (other than to test its functionality), but if you did, you could do so using this command.

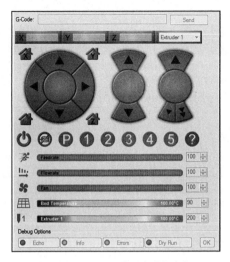

The Repetier manual control interface.

G-Code

In addition to these commands which usually have buttons, the host control software also has a way to send your own G-code commands. This is just a simple command line that allows you to send a command and receive a return (if applicable).

If you recall from earlier, G-code is what the 3D printer's control board actually uses for instructions. It is a long series of commands that call out a particular function along with coordinates, if necessary.

The commands are very basic, and the functionality comes from stringing many commands together. Each individual command is something simple, like trace a line from point A to point B or draw an arc with a specified radius between two points. But when these simple commands number in the thousands, they can produce very complex parts.

Printing with G-Code

When you use slicing software to slice a 3D model, that software's algorithms analyze the 3D model to determine what G-code commands it can use to reproduce the model. After the model has been sliced in thin layers, the slicer is then left with essentially a two-dimensional drawing, which it needs to trace with G-code.

HOT TIP

Technically speaking, 3D printers don't have to use G-code. They can use other instructions, if that's what they're programmed to use. In fact, some 3D printer manufacturers do build 3D printers that use proprietary languages for instructions. However, that's rare; most 3D printers today, especially in the consumer market, operate using G-code.

The software generates G-code based on rules determined by the values you set, along with the algorithms written by the programmer(s) of the software. Things like filament size, extrusion width, movement speed, and so on are all used to generate that G-code. Once a complete set of G-code instructions is generated, the G-code can then be fed to the 3D printer from the host software.

Performing Functions Manually with G-Code

Aside from when you're actually printing, you can also send G-code commands to perform functions manually, such as changing EEPROM values, auto-leveling the bed, checking the status of the end stops, and so on.

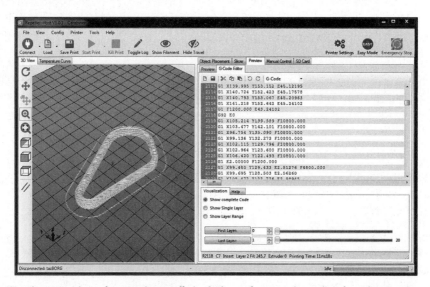

G-code commands can be entered manually in the host software and sent directly to the 3D printer.

G-code commands have a very simple syntax; all you have to do is enter the command name (for example, M212) followed by the required values, if necessary. If you're just requesting information, no additional values are needed most of the time. If the command allows you to enter multiple values, you can enter any combination of them. If, for instance, you were doing a homing command, you could specify if you wanted to home X, Y, Z, or any combination of the three.

There are hundreds of possible G-code commands, although not all of them will work with 3D printers (and some of them may even have different effects, depending on the particular firmware). While the vast majority of them aren't very useful or practical to use manually, the following are a handful of them that might prove to be helpful to you:

G20 (units to inches): You can use this command to change the current units from the (generally) default millimeters to inches.

G21 (units to mm): This sets the current units to millimeters, which is generally the default.

G28 (homing): This is used to home an axis or axes. If just "G28" is entered, all three axes will be homed. However, you can enter "G28 X Z," for example, and it will just home the X and Z axes and skip the Y axis.

G29 (auto-leveling): If your 3D printer has a probe, you can use this to auto-level the bed by probing three points. This uses the probe to measure the height of three corners of the bed in order to calculate the plane of the bed. This, in turn, can be used to raise and lower the Z height throughout the print to keep the nozzle at a constant distance from the bed.

G92 (setting coordinates): If you need to manually tell the 3D printer what its current coordinates are, you can use this command. G92 alone sets all three axes to zero. It can be appended with a number to set an axis at that coordinate without changing the others; for example, "X52.3" would set the X axis to 52.3 without changing the other axes. Any combination can be used to specify individual axes.

M92 (setting steps): If you need to change how many steps are needed per millimeter of movement—such as when you're calibrating your printer—you can do so with this command. For example, "M92 Y200" would set the steps per millimeter to 200 on the Y axis.

M119 (end stop status): This tells you if each of the end stops is currently reading high or low, which should let you know if they're triggered or not. You can try pushing an end stop switch with your finger when you run the command to test if the end stop is working (the state should change).

M500 (storing EEPROM values): If you have modified any of the EEPROM values (like steps per mm) using G-code commands, this command stores them permanently in the EEPROM. Be careful with this, as it's easy to accidentally overwrite your settings with unusable values.

M503 (reading EEPROM values): This command displays all of the values currently stored in the EEPROM. The command name at the beginning of each returned value can also be used to change the particular values on that line. So if a returned line reads "M92 X200.0 Y200.0 Z802.2," you can change any of those values by entering something like "M92 Z820.3"

While you may not ever have to actually send any manual G-code commands, it's a good idea to become at least vaguely familiar with these in case you ever need them.

WATCH OUT!

If you were so inclined, you could technically print an entire part by manually sending G-code commands to the 3D printer. Of course, doing so would be hugely impractical, and there really wouldn't be any point in doing so.

The Least You Need to Know

- Firmware is base-level software that runs on the control board and is stored in ROM. It's written specifically for a particular control board and hardware setup.
- Host software is what allows your computer to connect to your 3D printer. It also sends commands to run prints.
- Slicing software generates the G-code instructions that are sent to the 3D printer by the host software. It's often combined into a single package with the host software.
- G-code is the programming language used to issue commands to most 3D printers. This is what the slicing software generates to print parts, but you can also enter G-code commands manually to perform some useful functions.

CHAPTER 12

Leveling the Bed and Adjusting the Z Height

You've already learned how important it is for the first layer of a print to stick well to the bed. The material the build platform is made from is important, as are the surface treatments you apply to it. But the filament will never stick well to the bed if it's not level or if your Z height isn't set properly.

The Z height is how far the nozzle is from the print bed. It's important, because it determines how much the filament is pushed onto the bed as it is extruded. The bed being level is equally important, because otherwise the Z height will change depending on what area of the bed the nozzle is over. In this chapter, I go over how to get the bed and Z height where you need them to be.

In This Chapter

- Why your bed needs to be level
- Manually leveling or auto-leveling your bed
- How to set your Z height

Why Does Your Bed Need to Be Leveled?

When I say the bed needs to be level, I don't mean it needs to be level relative to Earth; you don't even want to touch a bubble level for this. Instead, the bed needs to be level relative to the axes of the 3D printer. Basically, the Z axis needs to be perpendicular to the build platform, and the X and Y axes need to be perfectly parallel to the build platform.

Why does it matter? A level bed ensures the nozzle is always a consistent height above the bed. The first layer of a print is very sensitive to how far the nozzle is from the bed. Too high, and the filament won't stick well; too low, and the nozzle will drag through the filament.

The acceptable height is a very small window (usually only one tenth of a millimeter or so). So even if the bed is only slightly off level, it could easily make parts of the bed too high or too low.

For example, imagine you had a 150mm-wide bed and had your Z height set at .25mm at one end. If the bed were just 1° off from level, the Z height at the other end of the bed would be more than 2.8mm. That could easily cause your print to fail, because the filament wouldn't adhere well at the far end.

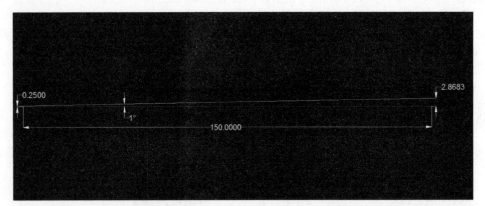

Even at just 1° from level, the print would be ruined.

This makes leveling virtually a necessity in order to print successfully. It should be the first thing you do when you first set up a 3D printer, and then fairly frequently to make sure the bed stays level. There are two ways to level a print bed: manual leveling and auto-leveling.

Manual Leveling

Most consumer 3D printers out right now require that you manually level the print bed. It's virtually impossible to manufacture a printer with a perfectly level bed that stays level, so most manufacturers include some sort of leveling mechanism in the design.

The build platform is generally attached to three or four screws with springs, and you tighten or loosen the screws in order to adjust the height of the bed at each point. Other mechanisms are used in some printers, but all of them have to provide some way of adjusting the height in at least three places in order to keep the bed level relative to both the X axis and the Y axis.

 HOT TIP

Leveling a build platform also requires it to be completely flat. If the bed is warped or bent at all, it will be impossible to level. There will always be part of the bed that isn't completely parallel to the X and Y axes. This is why glass and borosilicate are so commonly used for build platforms; those materials are easy to make and keep flat. Other materials, like aluminum, are still used for build platforms but can pretty easily become warped. Still, even with a perfectly flat build platform, the bed still has to be leveled.

To level the bed, all you'll need is a piece of paper. The idea here is that you'll set your Z height at one corner so the nozzle just barely touches the bed. You then place the piece of paper between the nozzle and bed and pull it out. Pay attention to how much force it takes to pull the paper out and how much resistance there is.

Manually leveling the build platform by using a piece of paper to measure resistance.

For the first corner, you'll just want a small amount of resistance that can be overcome easily with a little force (without tearing the paper). You then raise the nozzle and move it onto to another corner (preferably the one diagonally across the bed) and lower it on the paper there. Pull the paper out and pay attention to how much resistance it takes. If takes more resistance to pull the paper out on the second point than it did on the first, you'll need to lower that corner a bit. If it takes less resistance, you'll need to raise that corner. Fine-tune it until it takes approximately the same amount of resistance as the first corner.

Next, repeat the process for the third and fourth corners. The resistance it takes to pull the paper out should be equal between all four corners. Once all four corners have equal resistance, the bed will be level.

One thing you should keep in mind is that this process isn't quite as simple as it seems in theory. It takes some time to do and can be fairly frustrating. It might take you a few tries to get a sense for how much resistance you need to feel in the paper, and once you start printing, you may notice that filament adheres better in some places than others. If that's the case, you can just make some fine adjustments, like a quarter turn of the screws.

Auto-Leveling

As you can imagine, manually leveling the print bed isn't exactly a fun process. It can be time consuming and frustrating. For that reason, many 3D printer manufacturers have released printers that include auto-leveling.

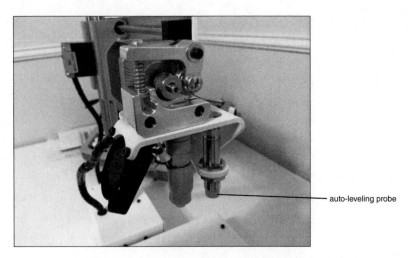

The auto-leveling probe on a Printrbot Simple.

So how does auto-leveling work? It doesn't physically adjust the bed at all. In fact, auto-leveling systems don't have any moving parts at all. Instead, they work completely in the firmware of the 3D printer.

Auto-leveling systems essentially work by virtually leveling the bed. They do this by raising and lowering the nozzle depending on what point of the bed it's at. They do this to keep the nozzle at a consistent height above the bed, no matter what point it's at.

For this system to work, the printer's firmware has to know how much to raise or lower the nozzle depending on where it is at in the X and Y directions. This is done with either a physical probe (basically just a switch), an inductive probe (which is triggered when it senses metal), or an optical probe (which is triggered when it gets close enough to see the bed). The firmware controls the process and moves the probe to three corners of the bed to measure their heights relative to each other.

FASCINATING FACT

The probes used in auto-leveling systems can either be triggered by physical contact or by sensing the bed without touching it. In this way, the probes work similarly to end stop switches.

Once the firmware has the height information for each of the three corners, it can use that information to determine how far off level the bed is. It basically constructs a flat plane from the three points, which represents the actual bed. It then knows how far to move the nozzle up and down at different points on the bed.

For example, if one corner is 1mm lower than another corner, the printer knows to lower the nozzle by 1mm when it's at that corner. Of course, it doesn't just suddenly lower or raise the nozzle when it reaches a corner. It will move the nozzle up or down gradually as it approaches each corner as necessary to keep the nozzle at a constant distance from the bed.

This makes auto-leveling a very convenient feature; however, it does have a couple of drawbacks. The first is that it relies on the accuracy of the probe, which may not always be perfect. The second is that it has no way to compensate for warped beds. For instance, if one corner of the bed was warped, you could still manually level the bed and just avoid printing in that corner, which can't happen when you use auto-leveling.

However, those are relatively minor drawbacks. For most people, the convenience of auto-leveling far outweighs the potential problems. And for that reason, auto-leveling is rapidly becoming more popular and common.

How to Adjust Your Z Height

Once your bed has been leveled, you can work on setting your Z height. The Z height is how far the tip of the nozzle is from the surface of the print bed. This will be just a fraction of a millimeter and needs to be set very precisely.

Unfortunately, setting it isn't just a matter of whipping out a ruler and measuring a specific distance. The necessary gap between the nozzle and the bed can vary dramatically depending on a number of factors, such as the following:

- Filament size
- Nozzle size
- Layer height
- Other print settings

All of those factors have one result: how much plastic is being squeezed out of the nozzle. That, in turn, determines how much of a gap needs to be between the nozzle and the bed. Technically, you could probably calculate what the ideal height is. But even if you did calculate it, it would be difficult to manually set it to that height.

So how do people go about actually setting the Z height then? It's mostly a matter of trial and error that goes like this:

- Set the nozzle close to the bed with just a tiny gap, and start printing.
- If the filament doesn't stick, lower the nozzle.
- If the filament is getting smashed into the bed by the nozzle, raise the nozzle.

You can go back and forth like this a few times until you get it right.

Setting the Z Height Manually or Automatically

On most 3D printers, you'll be setting the Z height manually. There is some sort of screw mechanism to raise or lower the Z stop (where the end stop switch makes contact), so you just make small adjustments with the screw to change the Z height.

However, on printers with auto-leveling, the probe usually acts as your end stop switch. In that case, either the probe itself is physically adjusted to change the Z height, or it's done in software. If it's done in software, the idea is to specify how much the nozzle needs to be raised back up after the probe is triggered.

Both systems work well, though manual adjustments usually tend to be quicker. That's because you don't have to fiddle with any settings in the firmware—all you have to do is make a quick turn on the adjustment screw.

Knowing the "Correct" Z Height

Knowing what the "correct" Z height is takes a little experience. As I said, you're looking for how the extruded filament sticks to the bed. If it's just barely clinging on (or worse, doesn't stick at all), it's obviously too high. This will usually look like a thread just lying on the bed. When the Z height is too high, it's obvious why there is going to be a problem: the filament just won't stick well.

Noticing when it's too low is a bit more difficult though. In that case, it will look like a flat line, almost like it's being drawn onto the bed. Unfortunately, the reason this is a problem isn't always a problem until later.

HOT TIP

Always pay close attention to your first layer. It's the most important part of the print, and is vital to the success of the print. Even if you've leveled your bed and set your Z height properly, it's possible that it could need to be recalibrated (it's normal for 3D printers to need recalibration occasionally). Watching the first layer is the best way to make sure it's correctly adjusted.

What happens is that the filament still has to go somewhere, so it tends to expand out. As each new line of filament is added next to the previous one, waves start to develop in the lines from the nozzle dragging through it. Eventually, this adds up and the nozzle will be bouncing all over the place on these waves, ruining your accuracy in the first few layers.

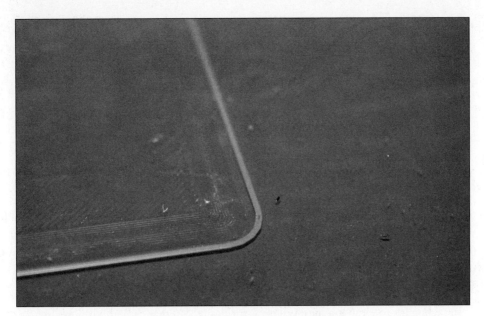

The first filament lines on the first layer should adhere well without be squished by the nozzle. As you can see with this layer, the nozzle was too low and squished the filament on the corner.

So what should it look like? You want a nice three-dimensional shape, but not so much so that it's just a thread of filament lying on the bed. If the filament naturally comes out in a circular cross-section, you're looking for it to have a sort of squished oval cross-section on the bed.

Learning what the perfect Z height looks like will take some experience. You'll need to see the effects of it being too high and being too low. Luckily, it doesn't take long to get it right. It should also be apparent pretty soon in the printing process if it's not right.

In the end, leveling the bed and setting the Z height are all about getting that first layer right. If you get that first layer right, you can usually count on the print turning out well.

The Least You Need to Know

- It's imperative that the bed be level in order to successfully print a part.
- Manually leveling the bed isn't a fun process, but it can be done with just a piece of paper.
- Many manufacturers are now including auto-leveling systems, which are more convenient than manual leveling. They tell the printer to either raise or lower the nozzle based on readings from a probe.
- Properly setting your Z height is as important as leveling the bed. It needs to be set so the filament adheres well without being smashed by the nozzle.

CHAPTER

13

Slicing and Printing

So you've got your printer all unpacked and set up. Your bed is leveled and your Z height is set up. You know practically everything there is to know about the history of 3D printing and how different kinds of 3D printers work. Now it's time to actually start printing!

To do that, you need to get your slicing software configured. You also need to learn about how to set up a new print and how to run it. In this chapter, I go over both these things. Don't worry, you're in the home stretch. Soon you'll be in 3D printing bliss!

In This Chapter

- Explanations of all of the slicer settings
- How to prepare your object for printing
- How to get a print started

Configuring Your Slicing Software

If you've followed Chapter 11, you have already set up your host software, and most likely the host software had slicing software bundled with it. (If you took my recommendation, you're using Slic3r, which is bundled with Repetier.) You have also entered 3D printer-specific settings in the host software. However, you still need to do the same for your slicing software.

Why? Because they're still separate pieces of software, even if they were bundled together. Technically, both pieces of software can be used independently of each other. So each has its own settings and needs to be configured individually.

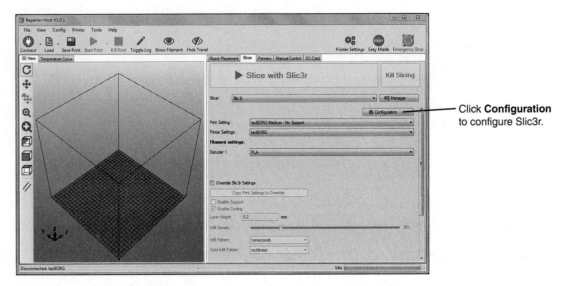

The Slic3r configuration can be found within the Repetier software.

If you're using Repetier, configuring Slic3r is pretty easy. From the **Slicer** tab in Repetier, you can just click the **Configuration** button to begin. If you're using another software combination, you can find similar settings within the particular slicing software you're using.

However, you shouldn't lament if you're using something other than Slic3r. It's all pretty much the same, and similar settings will be found in all slicing software (maybe with a slightly different name). For that reason, I'm now going to cover what each of the slicer settings actually does. Once you understand what the settings actually do, you'll have no trouble using them regardless of what they're called.

Slicer Settings Explained

In general, the slicing settings you'll be configuring can be roughly divided into three different types: printer settings, filament settings, and print settings. In the Slic3r configuration window, these are helpfully divided up into three separate tabs for your convenience.

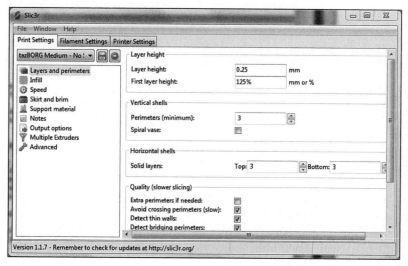

The settings in the Slic3r configuration window are divided into three types: print settings, filament settings, and printer settings. This shows the different tabs for them in the window.

In other slicing software, they might be divided up differently (or not divided up at all). Still, it's a good idea to think of them as separate kinds of settings. This will help you mentally organize them into those settings you'll need to tweak frequently, occasionally, or rarely.

 **HOT TIP**

Most software will allow you to save multiple configurations. This lets you quickly switch between predefined settings that you set up for specific types of prints, specific filaments, and even for multiple 3D printers. I advise you to take advantage of multiple configurations, so you can reuse settings without constantly tweaking them.

Printer Settings

The printer settings are the settings you should rarely have to modify. Ideally, you'll just set them up once and then never touch them again—or at least not until you get a new 3D printer. These tell the slicing software a little bit about the printer itself, so it knows what it's working with.

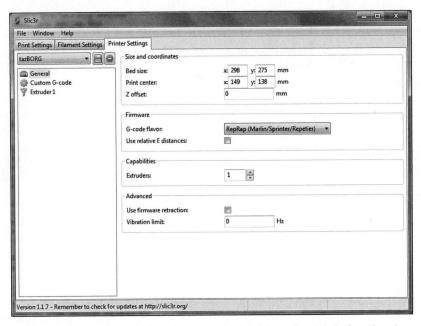

The printer settings, which you'll rarely change, give the slicing software info about the printer.

Printer settings are the first thing you should configure. Luckily, there isn't a whole lot to set up here. The following are the settings you'll encounter (keep in mind that these may be in separate sections of the Printer Settings tab):

Bed size: This is an easy one to start with! All you have to do is enter the size of your print bed in the X and Y directions.

Print center: As you might expect, this is just the X and Y coordinates for the center of the bed. Generally, you just divide the bed size by 2 and enter those numbers here. But you can modify this if you want it to center your prints in another spot.

G-code flavor: Here you'll specify what kind of G-code your printer uses. This might be a head-scratcher at first, but different control boards can actually interpret G-code slightly differently. If you don't see your specific printer or control board listed, it's usually safe to use "RepRap."

Use relative E distances: You can check this box if you want to use relative G-code distances (as opposed to absolute coordinates). Unless your manufacturer says you need to check this, just leave it unchecked.

Extruders: The number of extruders your 3D printer has can be entered here. Normally it'll be just 1, but if you have multiple extruders, you can specify that here. If you do, it will add sections to set up the additional extruders.

Use firmware retraction: This lets your firmware handle the retraction settings (as opposed to having your slicing software do it manually). In most cases, you'll want to leave this off unless you have a reason not to. However, if the software retraction settings don't work properly, you can use the firmware settings.

Vibration limit: You can attempt to reduce printer vibration problems with this setting. However, it's difficult to get it to work right, and the vibrations usually aren't substantial. I recommend you leave this at 0 to disable it.

Start G-code: Whatever you enter here will be added to the beginning of every G-code file the slicer outputs. So these G-code functions will be performed at the beginning of every print. Usually, people use this for things like homing all the axes (G28) or auto-leveling (G29).

End G-code: This works like the starting G-code, except it goes at the end (crazy, huh?). This is most often used to turn off everything on the printer, so it doesn't stay hot when you're away. For now, you can just leave the default commands there.

Layer change G-code: This also adds G-code, except it puts the G-code in between every layer. You can add code here that you want to be executed every time the layer changes (for instance, you could add a pause to let the previous layer cool before starting a new one).

Tool change G-code: G-code entered here will be inserted anytime the printer switches from using one extruder to using another. You can completely ignore this if you only have one extruder.

Nozzle diameter: This is where you need to enter the physical diameter of the nozzle opening on your hot end. If you don't know it, don't just guess! This is an important setting, so look up your nozzle diameter if you're not sure.

Position: If you have more than one extruder, this setting specifies how far apart the two nozzles of the extruders are.

Retraction length: Retraction reverses the extruder to pull the filament back out of the hot end slightly. This improves print quality by reducing blobs when line segments are completed. For this, 1mm or 2mm is usually a good setting to try.

Lift Z: This actually lifts the nozzle up (by the specified amount) when a line segment is finished and the printer is moving to another position. This can help keep the nozzle from dragging across your print.

Speed: This is just the speed the extruder will retract filament. Whatever the default is should work fine (usually 20 to 40 mm/s).

Extra length on restart: If you're having problems with filament not extruding immediately after a retraction, you can use this setting to extrude a little extra. However, you will have to experiment with the length to use.

Minimum travel after retraction: You can use this setting to skip retractions on moves that only cover a short distance. That distance can be adjusted to your liking.

Retract on layer change: This forces a retraction when you move up to the next layer of a print.

Wipe while retracting: To further reduce blobs and stringing, you can use this setting to physically wipe the nozzle on the print between moves. Personally, I recommend that you turn this off, because it has a tendency to mess up the surface as it wipes (as you might expect from dragging the hot nozzle on the plastic).

Filament Settings

These are settings you shouldn't have to modify often but that you will have to adjust occasionally. Basically, they're settings that define the specifications of your particular filament and how it should be used by the printer.

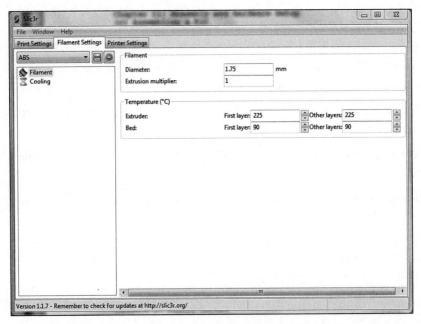

The filament settings are there for you to enter the specifications of the filament you're using.

You'll usually need to modify these whenever you put a new roll of filament on (at least slightly). This is because every material, manufacturer, color, and even specific roll is slightly different. The diameter of the filament, ideal extrusion temperature, bed temperature, and cooling settings will all vary.

The following are the different filament settings you'll deal with:

Diameter: This setting is easily one of the biggest causes of print quality problems. While you can just set it to 1.75mm or 3.00mm, the actual filament is rarely actually that size. Instead, for every roll of filament you get, you should measure the diameter in a few places with *calipers* and enter the average measurement here. This ensures that the slicing software accurately calculates how much filament to extrude.

 **DEFINITION**

Calipers are a common measurement tool used in a wide range of fields and industries. They come in both analog and digital varieties and are used to measure lengths, distances, and depths. Calipers are capable of very high precision and commonly come in 6- to 12-inch sizes (though larger and smaller ones exist).

Extrusion multiplier: You can use this setting to fine-tune how much filament is being extruded. For example, if you notice the printer is overextruding, you can set it to .95 so it only extrudes 95 percent of what it normally calculates is necessary.

First layer extruder temperature: This is where you set the temperature for the filament you're using. The first layer can be set independently because some people like to make the first layer hotter in order for it to stick better.

Other layers extruder temperature: This is the filament extrusion temperature for the rest of the layers (after the first one).

First layer bed temperature: If you have a heated bed, you can set its temperature for the first layer here.

Other layers bed temperature: This is where you set the bed temperature for the rest of the layers.

Keep fan always on: If you have a print fan you want to run all the time, check this. However, I recommend you don't do so for ABS, as it causes warping problems.

Enable auto cooling: This lets the slicing software decide when to turn the fan on, depending on what you specify in the other settings.

Fan speed: This is the minimum and maximum fan speed used at any time during the print. Different fans respond to this differently (for example, some won't even turn on at all below a certain number). Pay attention to how your fan acts and sounds, and fine-tune it with these settings.

Bridges fan speed: Bridges are parts of the print where the filament has to be extruded across an empty space with no other material below it. To keep these from drooping, it's recommended that you cool it quickly with a fan. So if you have a fan, enter a number to turn this up.

Disable fan for the first: As I mentioned previously, some people like their first layer to be hot to help it stick. With this setting, you can turn off the fan for a specified number of layers in the beginning to help with adhesion.

Enable fan if layer print time is below: If a particular layer of the print is very small, there may not be adequate time for it to cool before the next layer is added. You can use this setting to force the fan to come on to cool those small layers.

Slow down if layer print time is below: In extreme cases (very small layers), the fan isn't enough to cool the layer before the next is added. This can cause the layers to sort of melt and deform. This setting forces the printer to slow down on these layers to give them time to cool.

Min print speed: This is slowest the printer is allowed to move on those small layers. Setting this too low can make small parts take a very long time to print.

Print Settings

So you're at the final tab of the slicer settings. Do you want the good news or bad news first? Let's start with the bad news: this tab has the most settings by far. The good news is that a lot of these settings are for advanced usage. For the time being, it's best to just leave those at the defaults or automatic setting.

 HOT TIP

A lot of the print settings are for really fine-tuning specific areas of the print, meaning they're really for advanced users. So if you don't see a setting listed in this chapter, don't fret! Just leave it at the default or automatic setting and move on. You can always look up the specific setting if you're really curious about it.

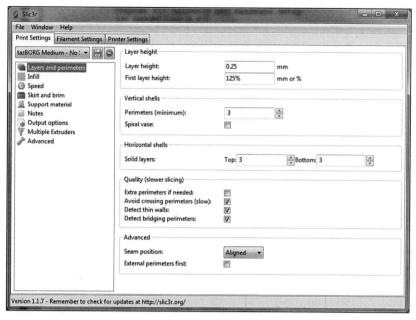

The Print Settings tab can be a scary place, but many of the settings can (and should) be left on their defaults.

The following are the various print settings you'll encounter:

Layer height: If you purchased your printer, the manufacturer probably gave you a recommendation for this setting. If not, you can safely set this to about 65 to 75 percent of your nozzle diameter.

First layer height: Often, it's good to make your first layer bigger to improve adhesion. A good number to start with is 125 percent.

Perimeters (minimum): In order to give the walls of the part a nice, solid structure, you need to specify a minimum number of solid perimeters. Two to four are good starting points.

Spiral vase: This nifty feature allows the entire part to be created from one continuous layer that slowly rises. However, it really only works with a single perimeter and no infill. So you should only use this for tall, hollow objects (like a vase), meaning you should normally leave this black or unchecked.

Solid layers (top): This is how many solid (noninfill) layers are on the top. It's similar to perimeters, just for horizontal walls. Again, 2 to 4 should work well.

Solid layers (bottom): This is same thing as the previous setting, just for the bottom layers.

Fill density: Infill is used on the interior of the part to avoid having to make it solid. This reduces weight, filament usage, and warping. For parts that need to be strong, use 50 to 80 percent. For parts that don't need to be strong, use 20 to 50 percent.

Fill pattern: This indicates what kind of pattern will be used for the infill. Honeycomb and rectilinear normally work the best (with rectilinear usually the fastest).

Top/bottom fill pattern: If you want a different pattern for the top and bottom layers, you can specify it here. However, I'd recommend sticking with rectilinear.

Speed settings: Unfortunately, I can't make recommendations for these settings, as they're very highly dependent on your specific printer. Start with the manufacturer's recommendations or with the defaults.

Skirt loops: In order to get the extruder flowing well before you actually start printing the part, you can have it print a couple of loops around the part. This has the dual purpose of allowing you to make sure it's completely on the print bed. You can just enter the number of loops you want the printer to make before it starts printing the actual part.

Distance from object: This is how far the loops will be printed from the edges of the part. This needs to be farther than the brim (if you use it), which I'll go over in a moment. If the part is very large and barely fits on the bed, you may need to reduce this to make the overall print smaller.

Skirt height: You can use this to make the skirt very tall to build a sort of wall around your part. This can help reduce drafts on the part in order to avoid warping.

Minimum extrusion length: You can set this if you want to make sure the skirt uses a certain amount of filament. Usually, you can just leave this at 0 unless you have extrusion issues (such as underextrusion in the beginning).

Brim width: A brim is a sort of base that's printed around your part. This can be useful for making sure the base of the part stays stuck to the bed and for keeping tall, thin parts upright. You can specify the width of the brim around the part, and the brim can easily be removed after the print is finished.

Support settings: If the part has overhangs and will require supports, you can turn on the support generation here. However, I recommend using the default and/or automatic settings for the best results.

Raft: A raft serves a similar purpose to the brim, except it's printed underneath the part. You can specify the thickness of the raft, as well as the interfacing layers with the part.

Multiple extruders: If you're using multiple extruders, this section is for you. Essentially, it allows you to specify which extruders are used for different operations. For example, you could use one extruder for supports and the other for the actual part.

Chapter 13: Slicing and Printing

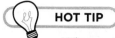 **HOT TIP**

When setting up multiple extruders, the settings that tend to be the most confusing are the offset settings. These are important, because they tell the slicing software where the second extruder is in relation to the first extruder. The nozzles of the two extruders must be at the same height, so this will specify how far the center of the second nozzle is from the first (in both the X and Y directions). If these are even a little off, the filament extruded by the second extruder will be deposited in the wrong place.

Preparing for and Running a Print

After you've put in your filament, printer, and print settings, it's time to get down to preparing for and running a print. This is where you'll spend most of your time when you're printing things. Because you'll frequently be modifying your print settings for specific kinds of parts, you'll always need to do the preparation and running steps to make sure your part comes out correctly.

Host Preparation

Preparing your print is a pretty simple but important process. I'll go over each specific step in more detail in the project sections of this book, but let me cover the general idea here.

I recommend that you begin by connecting to your printer and heating up the hot end and heated bed (if you have one). This will give it time to warm up while you're doing the rest of the steps. You can then load the part or parts you want to print.

With the parts loaded, you'll be able to view them in the Object Placement tab in Repetier. Here, you can move them around, rotate them, duplicate them, delete them, and so on. You should also take a moment to make sure everything fits on the bed and that they're all oriented properly. (Trying to print a part upside down will usually result in failure.)

If everything looks good, you can move over to the Slicer tab. If you're in Repetier and using Slic3r, make sure Slic3r is selected under the Slicer drop-down box. Below that, you'll notice drop-down menus for your Print Setting, Printer Setting, and Filament Setting configurations. Make sure the ones you're intending to use are selected. Once you've verified those, you can click the **Slice with Slic3r** button.

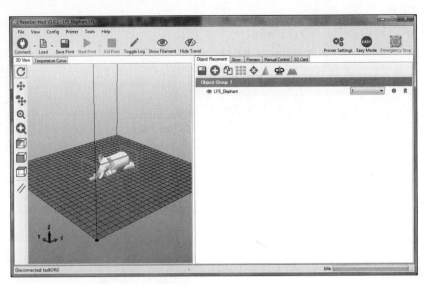

After loading your part(s), make sure they fit on your bed and are oriented properly. As you can see, this model of an elephant fits on the bed and is right side up.

Running a Print

Slicing your object or objects can take anywhere from a few seconds to hours depending on the complexity of the part, how powerful your computer is, and the settings you chose (for example, supports take more time to process). But once it's finished slicing, you should be presented with a preview of the print.

Look over the preview, which shows you where the slicing software is intending for the printer to actually extrude plastic. It should look like your original part but broken down into lines. If you're using supports, those will be shown as well. This is your chance to catch any serious setting problems that might cause the printer to do unexpected things.

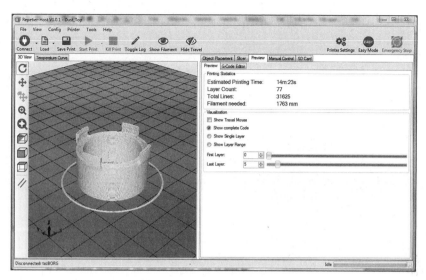

The preview shows the part to be printed broken down into lines.

If everything looks good, you can go ahead and push **Start Print** to start printing the part. Your printer must remain connected to the printer throughout the entire process (and it must be turned on, of course). It might take your printer a minute or two to start moving if it still needs to reach the required temperature.

Once it starts printing, don't go running off quite yet. Stick around for a few minutes while it prints the first layer. Your first layer is by far the most important one, because it's the foundation for the entire part. Make sure it sticks well to the bed and that the filament is extruding well. Then, in anywhere from a few minutes to a day or two, your part will be done!

The Least You Need to Know

- Most of the filament, printer, and print settings are important, so take the time to learn what they do. But when in doubt, stick with the defaults.
- Make sure your objects are placed and oriented correctly and that they fit on the bed before slicing.
- Always take a quick look at the preview to make sure everything looks right before starting a print.
- The first layer of a print is always the most important one. It's well worth your time to stick around for a few minutes to make sure that first layer goes well.

CHAPTER

14

Troubleshooting Your Prints

3D printing is pretty complicated, and that means there are a lot of problems you can run into. A few small tweaks in your settings can easily make the difference between a high-quality print and one that's going to get thrown away. But because there are so many factors, it can be difficult to figure out what the problem is. In this chapter, I walk you through some common issues and what you can do to fix them.

In This Chapter

- How to identify print problems
- Extrusion, temperature, and adhesion issues
- Fixing other common print quality issues

What's the Problem?

Obviously, there are a whole lot of things that can go wrong in the course of your 3D printing journey. If you're familiar with Murphy's Law, you know that anything that can go wrong will. These problems can be as minor as poor surface quality (like ripples) to major problems that cause your entire print to fail.

> **FASCINATING FACT**
>
> You can hold on to your failed prints for use in the future. They can be used to make ABS glue and even your own filament. If you do make your own filament, however, keep in mind that you shouldn't use more than 10 percent recycled plastic. If you don't have the ability to make filament yourself, you can drop the used ABS off at your local hackerspace or makerspace for them to use.

Most people will spend a great deal of time fine-tuning their settings to get really high-quality prints. But what if your parts are coming out terribly? It can be very frustrating if this is happening and you can't even achieve mediocre results.

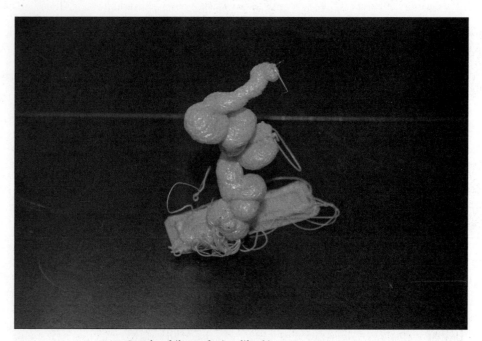

Complete failures of prints like this are not uncommon.

Luckily, these issues can usually be identified fairly easily by comparing your bad print to known problems. Once you've figured out what the primary cause of the problem is, you can adjust your settings to fix it. Actually getting the settings right might take a little trial and error, but at least you'll know where to focus your attention.

Extrusion Problems

Extrusion problems are those caused by improper filament extrusion. As you've learned multiple times in this book, extrusion is the key to FFF 3D printing. However, the slicing software has to make a lot of calculations based on many parameters in order to determine how much filament needs to be extruded. For the most part, the slicing software is good at these calculations, but with so many parameters, it's easy for one small error to result in a failed print.

Overextrusion

Overextrusion is when too much filament is pushed through the nozzle. It usually causes the surface of the print to look lumpy, uneven, and sloppy. This is almost always caused by a setting that was entered incorrectly though, in some very rare cases, it could be a hardware problem.

Overextrusion causes surface quality problems, like the bubbling around the octopus's eye.

The most common setting that causes this problem is the filament diameter. In order to fix this, you have to actually measure your filament's diameter (preferably in multiple places) in order to get an accurate number. Other settings that can cause this: the extrusion multiplier being set too high, the nozzle diameter being set too small, manual extrusion settings (try changing them to auto), and the extruder steps per mm (your extruder should be calibrated).

Underextrusion

The opposite of overextrusion is underextrusion, which is when too little filament is extruded. The effect of underextrusion is usually visible gaps between lines of filament. In extreme cases, this can cause a complete failure of the part, because layers don't adhere well enough to each other.

Underextrusion makes the filament strands too thin, causing gaps, as you can see in this printout of an octopus.

The causes of underextrusion are pretty much the same as overextrusion. It will just be in the opposite direction. For example, if you set your filament diameter too low, it will cause overextrusion. But, if you set it too high, it will cause underextrusion. The solution is simple: measure your filament's diameter and enter that.

Jamming

The most drastic (and obvious) extrusion problem is jamming or clogging. Either the printer will stop extruding altogether, intermittently stop extruding, or extrude thin and uneven lines of filament. This usually ends up completely ruining the part.

Jams usually happen when the filament gets twisted up in the cold end. This is easy enough to fix by just pulling it out, cleaning out the cold end, and starting over. Clogs in the nozzle are more difficult to deal with. They usually happen when the hot end temperature is too low, dust or dirt gets into the hot end, or the quality of the filament is low.

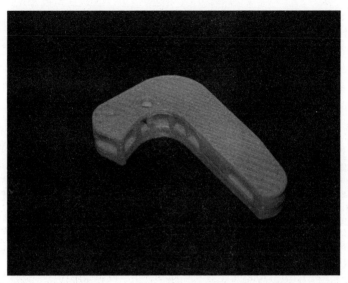

Jammed extruders and clogged nozzles can cause the printer to stop extruding completely. This will cause surfaces to be thin instead of solid, as you can see with this part.

 **WATCH OUT!**

Low-quality filament, especially in dark colors, is often difficult to melt and extrude properly. There is just too much nonplastic junk in the filament, and it causes the nozzle to clog.

When your nozzle becomes clogged with ABS, the best solution is to remove it and soak it in acetone in order to clean it out. For PLA or other materials, you can heat up the hot end and stick a piece of filament into it. Allow it to cool to about 150°C, and then pull the piece of filament out. You may need to repeat this a few times, but you can usually pull out the clog.

Poor Dimensional Accuracy

You may notice that your parts are coming out with the dimensions slightly off. Or you might try to fit two printed parts together only to find they don't fit. There are usually three possible culprits:

- Your steps aren't calibrated.
- Some part of your drive system is loose.
- The slicer isn't handling the slicing correctly.

When it comes to the first issue, each axis has to have the number of steps per millimeter calibrated. If it's not right, the lengths on that axis will be too short or too long. Most people notice this when circular features (like holes) end up coming out as ovals. That's because one axis is moving too far (or not far enough), causing the circle to become elongated. You can find the information for correct calibration from the manufacturer of your 3D printer or on community forums (see Appendix B).

In terms of the second issue, the drive system becoming loose can cause unintended movement, resulting in poor accuracy. Fixing this is usually a simple matter of tightening up all of your screws and making sure your belts are nice and taut.

And finally, the most frustrating cause is the slicing software itself. The shape of the filament as it's deposited is fairly difficult to account for mathematically, and many slicers don't quite get it right. This can make your part slightly too big (usually by a fraction of a millimeter). The only real solution to this is to try another slicer or to turn down your extrusion multiplier setting in order to slightly underextrude and make the part smaller.

Temperature Problems

The temperatures of the hot end and heated bed both play important roles in the quality of the printed part. Pretty obvious, right? But it's not always obvious when a temperature problem is the culprit for a particular problem, so I'm going to cover what common temperature-related problems look like.

Hot End Is Too Hot

The filament has to be melted just enough to flow freely out of the nozzle but still has to be able to quickly cool once it's been deposited. If the hot end is too hot, the filament won't cool quickly enough. The effect is that the part tends to look slightly melted (though it may only look this way in certain spots). The same problem can occur if the layer is small, because it doesn't have time to cool before a new layer is added.

If it's consistent (not just on small layers), you can turn your hot end temperature down. If it's just on small layers, you can use a fan for cooling or slow down small layers in your settings.

Chapter 14: Troubleshooting Your Prints 167

If the hot end is too hot or the layers are small without adequate cooling, the part will develop a melted appearance like the top of this one.

 HOT TIP

Extreme hot end temperature problems are often caused by problems with the thermistor. This is usually either because the thermistor has failed or has come loose. If it's loose, it can be affixed with Kapton tape. If that's not the problem, the thermistor can be replaced. (They're usually only a few dollars.) Contact the manufacturer for information on replacing the thermistor and to see if it will be covered under a warranty.

Hot End Is Too Cold

If the hot end is too cold, it tends to produce results similar to nozzle clogs. Either filament won't extrude at all or it will extrude inconsistently. All you have to do is simply turn the temperature of the hot end up.

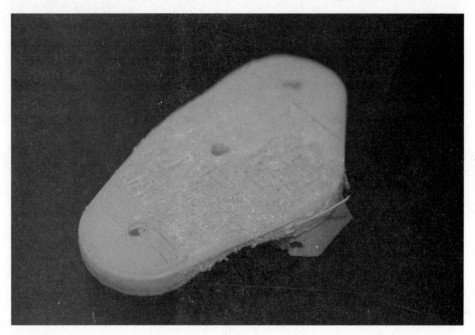

When the hot end temperature is too low, it results in inconsistent extrusion. This part shows how messy this issue can look.

Cracking of Part Due to Cooling

A part cracking is a common problem when you print large parts in ABS. This is because the ABS cools and contracts, splitting the part horizontally between layers. While this is a difficult problem to overcome, there are a few potential ways you can try to avoid this.

Because print fans and even drafts in the room can cool a part and cause contractions that lead to cracks, one solution is to keep the part from cooling off as much as possible. Enclosures (especially heated enclosures) are the best way to overcome this. However, if that's not feasible for you, you can try using a tall skirt to act as a shield around the print. Another option to avoid cracks is to design the part itself so it doesn't have any long, tall, horizontal surfaces in order to reduce the stress from contraction.

The contraction of the plastic as it cools can cause a part to crack horizontally between the layers.

Discoloration of Filament Due to Heat

If you're using a light-colored filament (especially white), you may notice the part becomes discolored. This is caused by heat, either from the heated bed or the hot end.

If it's just on the bottom of the part, it's the heated bed; if it's everywhere, it's the hot end. In both cases, it could signal the temperature is dramatically too high, meaning you should turn down the temperature. Sometimes with the hot end, however, your extrusion speed may be too slow and the filament could be sitting in the hot end too long. If that's the case, try speeding up your prints.

Adhesion Problems

You've learned in previous chapters how important it is for the filament to stick well to the bed. Those times when it doesn't can be a big headache for you. For most people, adhesion problems are the biggest source of frustration in 3D printing. This is because there are so many potential causes and so many potential solutions. Here, you learn what the results of poor adhesion are and what to do about them.

Warping

The same contraction that causes cracking also causes the part to warp. This is usually made evident when the bottom of the part curves (especially at the edges). This can be solved the same way as cracking—enclosures and designs that limit long, horizontal surfaces—but you can also dramatically reduce it by increasing bed adhesion.

The contraction on this ABS part caused extreme warping.

To increase adhesion between the first layer of filament and the bed, take a look at the surface treatment options mentioned in Chapter 8. Heated beds help a lot, as well as tapes, films, glue, and even hairspray.

Part Comes Loose

If your surface adhesion is really poor, the entire part might come off the bed during the print. Usually, this happens when your Z height is too high, or when you don't have any surface treatments at all. If you're using the proper surface treatments, try lowering your Z height a little bit to fix this issue.

FASCINATING FACT

Auto-leveling systems do a very good job of alleviating problems caused by improperly adjusted Z heights, because they allow you to set exactly how high it should be in the software. Even if your 3D printer didn't come with auto-leveling, it's possible that it could be available as an upgrade. If you're having frequent problems with setting the Z height on your printer, it might be worth checking to see if that's an option available to you.

Part Breaks During Removal

Sometimes your part might actually stick too well. This can cause the part to break during removal or even to break glass beds in extreme cases.

This particular problem is often caused by overuse of ABS juice/glue—a case of going a little overboard with a surface treatment. You should only use a thin layer of ABS juice to avoid this. However, good adhesion isn't a bad thing, so another option is to try using a flat tool to pry the part off without breaking it.

Other Printer Problems

Some problems just don't fit neatly into a predefined category. These are the issues that tend to come out of left field and often throw you for a loop. Most of these won't result in a complete failure of the print, but they're still irritating and affect the quality of the part.

Blobs

If you notice little blobs of plastic on the surface of the part, it's usually caused by either lack of retraction, overextrusion, or a hot end that's set too high.

I've already covered solutions for the latter two causes, and the first one is easy enough to fix in your settings. Simply turn on retraction in your slicer settings (1mm to 2mm usually does the trick).

Stringing

Stringing looks like you'd expect: thin strands of filament between features on the part. The causes for this are lack of retraction or the hot end's temperature being set too high. This can be solved by turning on retraction or lowering the temperature of the hot end.

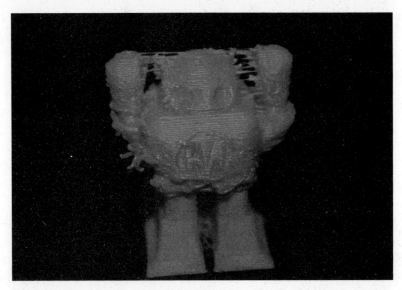

Turning on retraction will usually fix stringing problems like what's shown on this robot.

Drooping

Bridges are features where a new layer is placed over empty space. For these features, the printer has to extrude filament across the chasm from one solid to another. Ideally, these should turn out nice and flat. But if you notice they're drooping, it's usually caused by a lack of cooling. You might also see this same problem happen on other overhanging features.

Using a print fan will do a lot to reduce drooping on bridges and overhangs similar to this.

To fix this, you need to make sure your print fan is turned on. If you don't have a print fan, you'll have to add one. If the fan is on but the problem still occurs, you should make sure it's turned all the way up. If it's still occurring even after that, you should look into printable fan shrouds for your 3D printer, which more effectively direct the air from the fan to the part.

Ghosting

Ghosting is when faint lines appear on the surface offset around features, something usually only visible on fairly flat surfaces that are next to perpendicular features. It's caused by vibrations from the weight of the extruder carriage as it's moving back and forth.

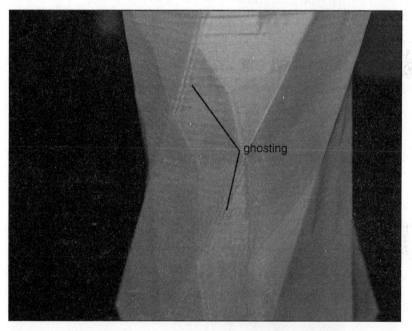

Ghosting like you see on this pencil holder can usually only be fixed by reducing print speed and/or acceleration.

The only way to fix this without modifying your printer's hardware is to slow down the print speed or to experiment with the acceleration settings. The acceleration settings might be hidden in an advanced area of your slicer's settings, but they determine how fast the printer can speed up and slow down as it moves an axis. Lowering the acceleration will allow the printer to still move quickly on long stretches but will make the changes in direction less abrupt.

The Least You Need to Know

- Incorrect settings can easily introduce many different kinds of print quality problems.
- Most print quality problems can be fixed by tweaking the print settings.
- In extreme cases, bad print settings can result in a complete failure of the print.

PROJECT 1

Carabiner

Project Time: 30 minutes

In this project, you'll be getting hands on and printing your very first part. For your first print, let's start with something quick and easy: a one-piece carabiner. This model should print quickly, and most printers will have it finished in under 20 minutes.

Before you begin, make sure you've downloaded the model pack from the Idiot's Guides website (idiotsguides.com/3dprinting) and have unzipped the folder so you can access the files. For this project and the subsequent projects, I'll be using Repetier Host and Slic3r, because both are free and popular programs. However, if you'd prefer to use other software (or your printer requires it), the steps should still be pretty similar.

Preheat the Extruder and Heated Bed

Start by opening Repetier (or whatever host software you're using) and connecting to your 3D printer. Once the connection has been made between your host software and your printer, click on the **Manual Control** tab to get to the temperature settings for the extruder and heated bed, which you'll now begin preheating. If you're printing in ABS, the extruder should be set at about 230°C and the heated bed should be somewhere between 80°C to 100°C. For PLA, the extruder should be 200°C and the heated bed (if you have one) should be between 50°C to 70°C.

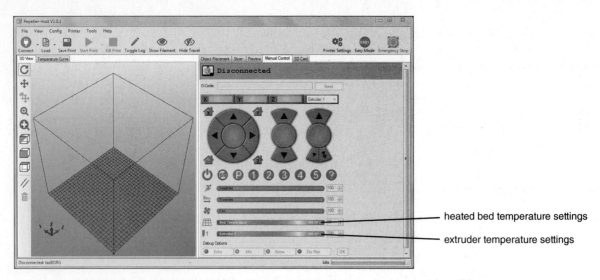

Preheat the extruder and heated bed to the appropriate temperature for the material you'll be using.

Load the .STL File

While the extruder and heated bed are heating up, go ahead and load the .STL file for the carabiner and slice it (found at idiotsguides.com/3dprinting). To do this, look for a button in your host software that says Load or Open. Push that button, and then find and select the **carabiner.stl** file. Slic3r, which runs within Repetier, will display the 3D model in the Object Placement window.

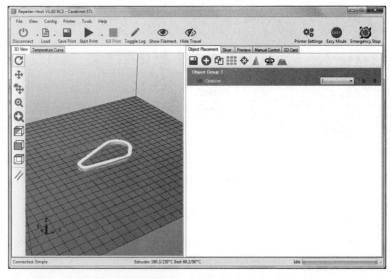

The Object Placement tab will display the loaded 3D model.

The gray box in the viewer window corresponds to your 3D printer's print area and is based on the settings you entered when you set up the host software. The carabiner is a small part that is only a few inches long, so it should be printable on virtually every 3D printer on the market.

 **WATCH OUT!**

If the model shows up very large (compared to the gray box), either your printer settings for the print area are wrong or you have your units set to inches (it should be in millimeters).

Slice the Model

With the model loaded, you can go ahead and slice it. With a simple part like this, the particular settings you use won't be crucial. However, make sure you have at least two or three perimeters (including top and bottom) and at least a 50 percent infill. This can be modified by going to the **Slicer** tab and clicking on **Configuration**. Once there, click the **Print Settings** tab and fill in the information if necessary under the **Layers and Perimeters** and **Infill** subsections. Your print temperature settings should be the same as what you preheated the extruder and heated bed to.

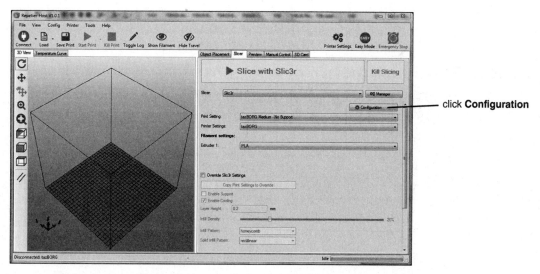

Check to be sure you have at least two or three perimeters (including top and bottom) and at least a 50 percent infill.

With your settings entered, you can start slicing the model by clicking **Slice with Slic3r**. How long it takes to slice depends on your computer and the slicer you're using, but it shouldn't take more than a few minutes.

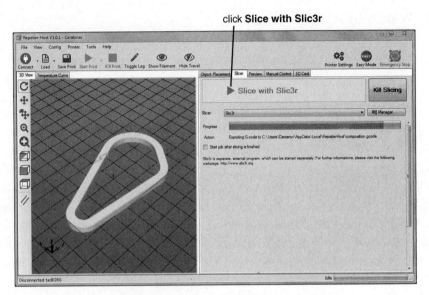

Slicing a small model like this should only take a few minutes.

Load the Filament

A lot of people keep the filament loaded all the time and only change it when they run out of filament or switch to a different material or color. But if you haven't loaded any filament yet, load it into the extruder now, while the slicer is running.

How you load the filament depends on the specific extruder your printer is equipped with. Some of them, like the Greg's Wade Reloaded I'm using, have thumb screws and a latch that needs to be opened up. Others have a lever that needs to be pushed to release tension on the bearing. Or it could be a completely different mechanism altogether.

However, no matter what kind of mechanism it is, the process is about the same once you have it open:

1. Trim the end of your filament with scissors at an angle, so there is a point (this helps it slide into the hot end).

2. Push the filament into the hole below the hobbled bolt or drive gear. With the hot end heated up, you should be able to push it in until melted filament starts coming out of the nozzle.

Load the filament by opening the mechanism and sliding the filament into the hot end.

Extrude Some Filament

Once you've closed the mechanism, switch back over to the **Manual Control** tab in your host software. There should be a button to extrude a specified amount of filament, which you can see in the following figure. Extrude 10mm or 50mm of filament in order to verify the filament has been loaded properly and to build up some pressure in the hot end (which helps improve flow while you're printing). This isn't absolutely necessary to do, but it helps to make sure everything is working.

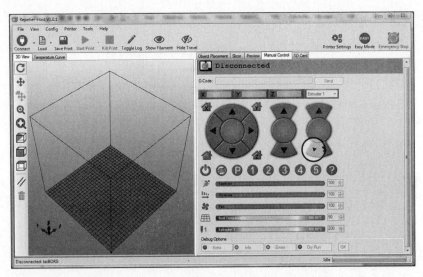

Extrude a small amount of filament to make sure it's loaded properly and to build pressure in the hot end.

When you extrude, you should have filament come out of the nozzle steadily in a consistent size. If it's erratic or comes out at an angle, something may be wrong. (If you run into this or any other problems, be sure to check Chapter 15, which covers troubleshooting.) Once you've determined that the filament is extruding correctly, you can pull off the pile of filament with a pair of pliers, being careful not to burn yourself on the hot end.

Make sure that the filament extrudes consistently, and then remove the plastic.

Start the Print

Now it's time to go ahead and start the print! Go over to the **Preview** tab, which gives you estimates on the print time and how much filament is needed (in millimeters). The viewer window shows you the G-code paths the slicer generated, so you can get an idea of what the printer will actually be doing from layer to layer.

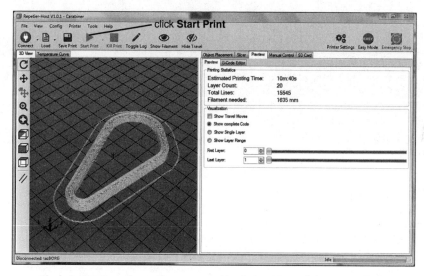

Check the Preview tab to make sure everything looks right, and then start printing.

If everything appears to be correct, push **Start Print**. Depending on the software and settings, the printer may not do anything immediately. Sometimes it will wait to make sure the temperature is steady before it begins. Once that's done, though, it should home each of the axes and start printing.

Watch the First Layer

Pay very close attention to the first layer being printed. The first layer is the most important one, because it's the foundation for the rest of the part. If the first layer is right, the chances are good the part will turn out right. You're looking to make sure you have good adhesion, without squishing the filament too much. You also want to make sure the flow of the extrusion is steady and consistent.

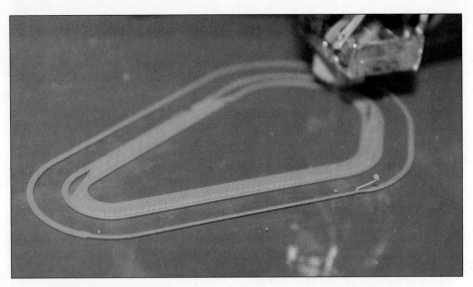

Pay close attention to the first layer, because it's the most important one.

 HOT TIP

I have my slicer set up to do a couple of loops around the part before it actually starts printing the first layer of the part. The purpose of this is to get the hot end flowing before it actually gets to the part, so I can see that the filament is adhering nicely all the way around (meaning the bed is level) and that the part actually fits on the bed. I recommend doing this for your prints as well.

After you've ensured the first layer is good, you can sit back and relax. However, it's still a good idea to check on the print every now and then to make sure there are no problems.

Let the Part Cool

Ten to 20 minutes later, your carabiner should be done! I know it's tempting, but resist the urge to immediately yank off the part. You want to give it five minutes or so to finish cooling off to make sure the plastic has finished hardening.

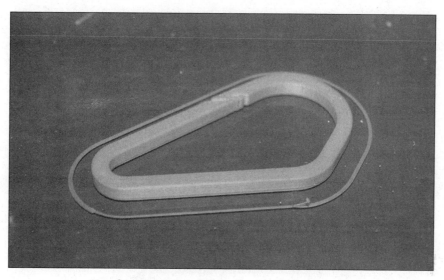

Your first print is done. Take a minute to admire your work!

Remove the Part

Once you've tested your patience by letting the part cool off, it's time to remove it. If you can't get a grip on the part, you can use a flat tool to pry it up from the bottom. Be careful, though; it's easy to break the part if the adhesion is good. You may need to work at it from different sides to get it off the bed.

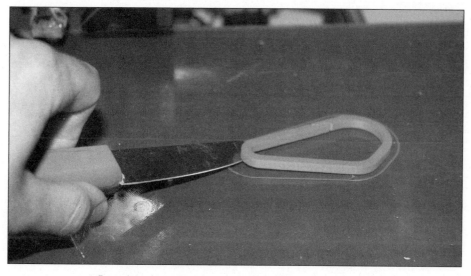

A flat tool can be used gently to remove the part if it's stuck on the bed.

If you're having a really hard time getting it off, give it some more time to cool down. Once the bed and the part have cooled down to room temperature, it should be a lot easier to remove.

Enjoy your first print, and spend some time contemplating the fact you just had a robot do your bidding to create a physical object from scratch!

Congratulations on your first part! The future is now!

PROJECT 2

Pencil Holder

Project Time: 4 hours

In this project, you'll be stepping up the difficulty a bit. Printing a pencil holder requires a couple of tricks to accomplish. As you'll notice once you open the file, the model is solid and doesn't open like a cup. In order to use it as a pencil holder, the Slic3r settings must be modified to remove the top and make the inside hollow. This should give you an idea of how powerful the slicing software is and how it can be used to customize the model.

Load the .STL File and Resize the Model, If Necessary

In Repetier, click **Load** and select the **pencil_holder.stl** file, which is included in the model pack at the Idiot's Guides website (idiotsguides.com/3dprinting). Next, check its size on the Object Placement tab, which should already be up. This model is a lot larger than the last one, so you should make sure it fits on your 3D printer.

If it doesn't fit, you can try resizing the model. On the **Object Placement** tab, you should be able to select the loaded model. You have the option to do basic modification, like repositioning, rotating, and resizing the model.

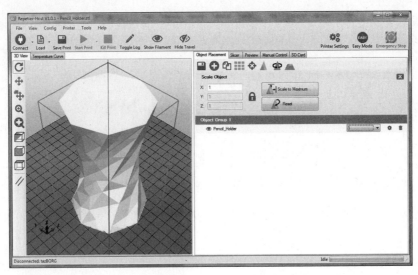

You can resize the model, if necessary, in the Object Placement tab.

While you can resize to fit on your printer, keep in mind that it can't be too small. Otherwise, it won't be very useful as a pencil holder.

Preheat the Extruder and Heated Bed, and Load the Filament

Make sure Repetier is connected to your 3D printer. Just like in project 1, you may want to start by getting the extruder and heated bed warmed up. If you recall, ABS requires settings of 230°C for the extruder and between 80°C to 100°C for the heated bed, while PLA requires settings of 200°C for the extruder and between 50°C to 70°C for the heated bed (if you have one). Both ABS and PLA work well for this model, so you should be fine using whichever you have on hand or prefer.

Once the extruder is up to temperature, load the filament into the hot end, if you haven't already. The steps for doing this are exactly the same as in project 1, so check that out if you need a refresher.

Modify the Slicer Settings to Make the Model a Cup

For this model, you have to make a couple of changes to the Slic3r settings in order to remove the top of the model and make it hollow; otherwise, it won't function very well as a pencil holder.

In Repetier, go to the **Slicer** tab and click the **Configuration** button. (See project 1 if you need a refresher.) Once the configuration window opens, under the **Print Settings** tab, locate and click on **Layers and Perimeters**. Here, you want to change the number of horizontal and bottom perimeters to 4. Then change the number of top perimeters to 0. This gives the bottom and sides of the model a solid thickness of 4 layers and removes the top of the model entirely.

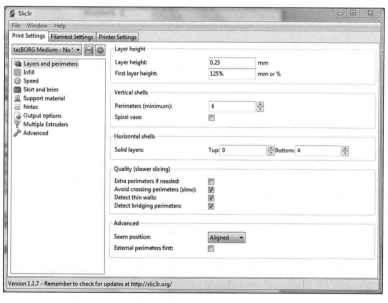

Turn off the top perimeters.

Next, click on **Infill**, which is located below Layers and Perimeters, and change the fill density to **0%**. This makes the inside of the model hollow.

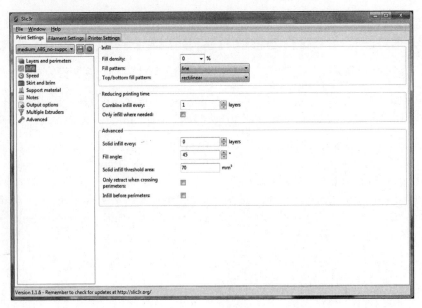

Set the fill density to 0%.

Slice the Model

With your Slic3r settings modified to make the model cuplike, you can go ahead and slice the model. Go to the **Slicer** tab and click **Slice with Slic3r**. This print should take 2 to 4 hours at the standard size, depending mostly on how fast your printer moves.

The actual slicing process might take a few minutes, so be patient here, especially if your computer is a bit older.

Start the Print

After the slicing has been completed, click on the **Preview** tab. You should see that the projected extrusion paths make a pencil holder instead of just a solid model. This is because the preview window shows you what is supposed to be printed, not the original model.

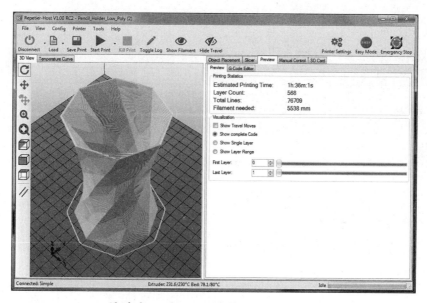

Check the preview to verify if the print looks right.

If you're happy with preview and the extruder and heated bed are up to temperature, you can start printing by clicking **Start Print**.

Watch the First Layer

As you should do with every print, stick around to watch the first layer. The first layer is the most important, so make sure it turns out right. You want good adhesion, but you don't want the hot end so close to the bed that it squishes the plastic.

Printing the first layer. Always check the first layer to make sure it looks correct.

Remove the Part

Once the print has finished, give it at least five minutes to cool off, so it can completely harden. Letting the heated bed cool down to room temperature makes it more likely that the part will just pop right off. But if you're eager to start storing your pencils in style, you can use a tool to remove the pencil holder. Just be sure to do this gently to avoid damaging the bed or the part.

You can use a tool to carefully remove the part from the bed.

With the part removed, you can admire your handiwork. To test it and make sure it works, place some pencils (or even pens) in it. If they don't fall out, then congratulations, you have a pencil holder!

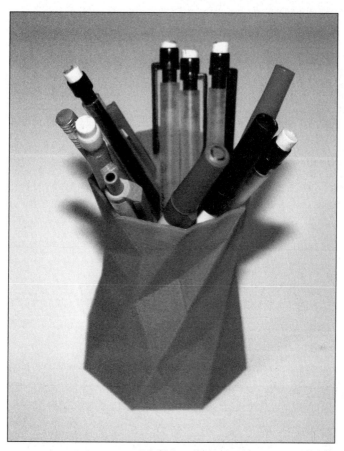

It holds pencils!

PROJECT

3

Robot

Project Time: 4 hours

For this project, you're going to be printing a robot figurine modeled by Nathan Deas. What makes this a bit trickier than the previous projects is the print will require the use of supports. This model has a lot of overhanging features, so removable support material has to be used to hold up those features during printing. After the print has finished, however, you can remove the supports.

Load the .STL File

As usual, start by loading the model. In Repetier, click **Load** and select the **robot.stl** file, which is included in the model pack at the Idiot's Guides website (idiotsguides.com/3dprinting). The model should load with the robot's back facing down, but if it's not, you can rotate it using the circular rotate button in the top-left corner.

HOT TIP

You also have the option of printing the model with the feet down. Printing it with the feet down will reduce the amount of the model that comes into contact with the support material, which should improve the surface quality of the overall print. However, because the feet are small, it will be less stable. To combat this, you'd want to print with a thick brim in order to ensure the model stays upright throughout the entire print.

Preheat the Extruder and Heated Bed, and Load the Filament

With the model loaded, it's a good idea to set the extruder and heated bed to the temperatures you'll use when you print. This model should turn out well in both ABS and PLA, so feel free to use either one. Just keep in mind that the extruder and heated bed need to be set at different temperatures depending on which material you use (see project 1 for the temperatures necessary for ABS versus PLA).

If your filament isn't already loaded, you should load it once the extruder is up to temperature; project 1 can give a refresher on the steps to do this, if you need them. If you're changing filament materials or colors, be sure to extrude at least 100mm of filament. Doing this will basically flush out the old filament.

Modify the Slicer Settings for Supports

Because this model requires supports, you need to change a few settings from the previous prints. The most important of these is turning on support generation. This tells the slicer to create supports where necessary.

Depending on the particular slicer you're using, the support settings may vary. If there is a simple automatic setting, start with that. Otherwise, you want to tell it to generate supports for overhangs with an angle over 45°. For now, leave the rest of the support settings at their defaults. Getting good supports takes a lot of experimentation, so start with something simple and then refine the settings on future prints.

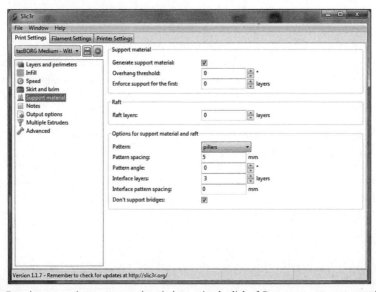

In Repetier, generating support settings is just a simple click of **Generate support material***.*

Other than the support settings, you also want to make sure your filament diameter is set correctly in order to make sure the supports connect just well enough to the part. Filament is rarely exactly the right diameter, so it's best to use a pair of digital calipers to measure. Even better, measure in three or four different places on the filament spool and then average the numbers. You can then change your filament diameter setting by clicking **Configuration** and then going to the **Filament Settings** tab to match the actual diameter of the filament.

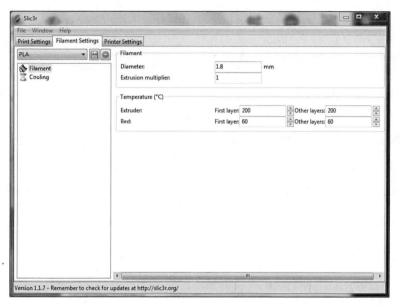

The Filament subsection of the Filament Settings tab has a place where you can adjust the filament diameter.

Also, if you're using PLA, make sure you turn on your print fan. You'll find the settings to turn on the fan in the Cooling subsection of the **Filament Settings** tab. This is always a good idea when printing PLA, but it's especially important when using support material. Without a fan running, support material might stick too well to the model when printing in PLA.

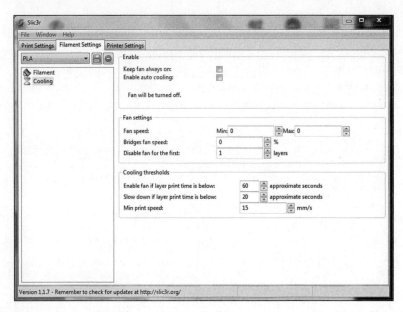

The Cooling subsection of the Filament Settings tab allows you to turn on and adjust the fan settings.

Slice the Model

Once you've updated your settings, you can slice the model. Go to the **Slicer** tab and click **Slice with Slic3r**. This particular model could take anywhere from 2 to 4 hours to print at its default size.

If you don't want to wait that long (who does?), you can scale down the model to something smaller. However, if you do scale the model, keep in mind that smaller features may not show up well on some printers. Luckily, this figurine doesn't have many fine features, so you should be able to scale it to 50 percent or so without any issues. The eyes may not show up well, but everything else should.

Start the Print

After the slicing has been completed, click on the **Preview** tab. You should see the model of the robot on its back, as well as the supports. If you're satisfied with the settings, support generation, and print time, go ahead and click **Start Print**.

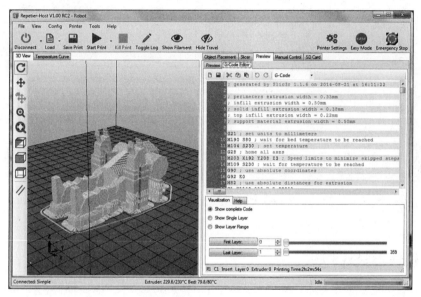

After slicing the robot model, the preview window should show you where supports will be generated.

Watch the First Layer

As always, pay close attention to the first layer to make sure you're getting good adhesion between the filament and the bed.

The first layer for this print may look a little different than the previous two. I'm using a brim on this print to increase the surface area of the first layer to help with adhesion, which is why there is a large, flat area around the model. (For more information on brims, check out the "Print Settings" section of Chapter 13.) But you may also notice a lot of small, jagged lines of filament around. These lines are the base of the support material the slicer generated, which is normal.

With this first layer, you can see the base of the support material.

Remove the Part

Once the print has been completed, you may notice it doesn't really look like a robot. This is because the model is partially surrounded by support material, which has to be removed. But first, the print has to be removed from the bed.

 **WATCH OUT!**

While the full-size model should be done after roughly three hours, you should check on how it's doing once or twice every hour. Especially for long prints like this, you don't want to wait until the end to discover there was some problem.

The finished print with supports should look something like this.

You can use your hands or a tool to remove the part from the bed, which should be easier to do once the part has cooled. Because there is so much support material, don't be too concerned if parts of it break off. The support material is designed to do that, so it's not a big deal (as long as the actual model stays intact).

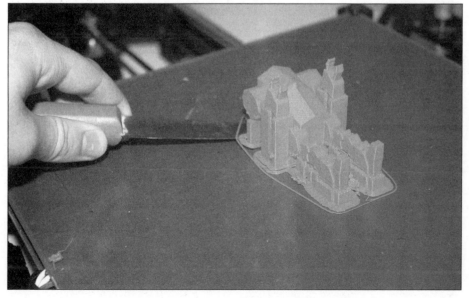

Wait for the print to cool, and then remove it from the bed.

Remove the Supports

After you've wrestled the print loose from the bed, it's time to go through the somewhat tedious process of removing the supports. The supports are connected to the model, but the slicer attempts to make that connection as fragile as possible. Ideally, the supports will just pop right off. But unfortunately, the reality isn't usually quite so simple. Instead, you'll probably need to use some pliers to grab the supports and work them off.

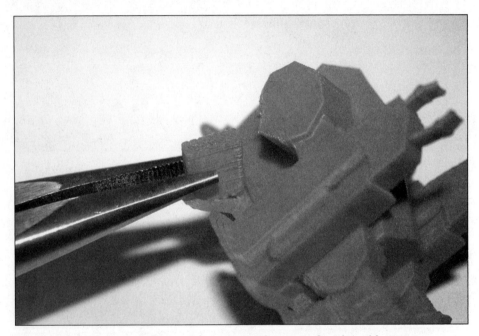

A pair of pliers can help you remove the supports from the model.

It can take some time to remove all of the support material from the model. Be careful not to damage the model, and resist the urge to use a hammer and chisel. Be patient and take your time here. Eventually, you should be able to pull off all of the support material.

The robot figurine with all of the supports removed.

 HOT TIP

With the supports removed, you may notice that the surface quality isn't nearly as good in the areas where the supports contacted the model. This can be improved by fine-tuning your settings, but there will always be some loss of quality here. For this reason, designers generally try to reduce the dependence on the supports.

Now enjoy your freshly printed robot, which was printed by a robot! Maybe you can print a whole robot army to defend your cubicle from invading coworkers?

The finished robot figurine, ready to follow your orders.

PROJECT

4

Storage Box with Drawers

Project Time: 1 day

It will take quite some time to print all of the parts for this project. That's because you're going to be printing four parts in total: the body of the storage box and three drawers. Each of these parts will take a few hours to print, so printing them is going to be an all-day affair.

But it will all be worth it in the end. This handy storage solution will give your nuts, bolts, and any other tiny items a comfy new home.

 HOT TIP

In a later project, you'll learn how to design and print your own custom drawers, so you'll be able to store whatever your heart desires (so long as it's small).

Open Your Host Software and Load the Storage Body .STL File

Start by opening up your printer host software. As usual, I'll be using Repetier for these photos. But if you're using other software, the process should be pretty similar. You can also preheat the hot end and heated bed, if you have one (project 1 provides a refresher on the temperatures necessary for ABS versus PLA, if you need it).

Next, click **Load** and select the **storage_body.stl** file, which is included in the model pack at the Idiot's Guides website (idiotsguides.com/3dprinting). The storage body should be lying on its end by default, but if it's not, you can use the rotation tools in the **Object Placement** tab to orient it properly.

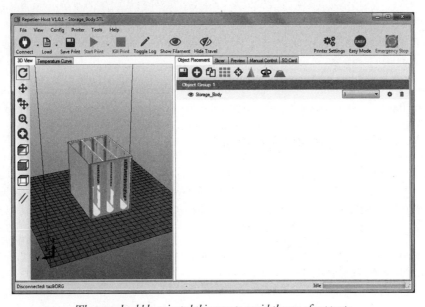

The part should be oriented this way to avoid the use of supports.

The purpose of orienting the model this way is to avoid having to use supports. The slots in the sides can be printed without supports because the ends are circular. Overhangs that are more vertical than 45° can usually be printed without supports. Because circles never get less than that, you can safely assume they'll print without supports.

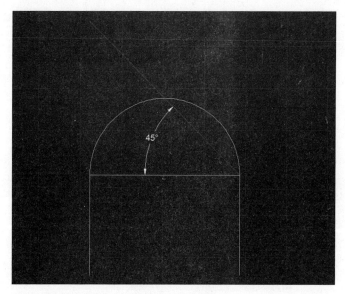

Circular features, like holes and slots, can be printed without supports. This is because the angle is never less than 45°, as you can see here.

Slice the Storage Body

Now it's time to slice the model. This model won't require supports or any special settings. However, you should make sure the infill is set fairly high, at least 50 percent (see project 2 if you need a reminder of how to set the infill percentage). In order to reduce warping, I also recommend you print this in PLA and with a brim.

Once you've got your model oriented properly and your settings tweaked, go to the **Slicer** tab and click **Slice with Slic3r**. This could take a while, especially on slower computers.

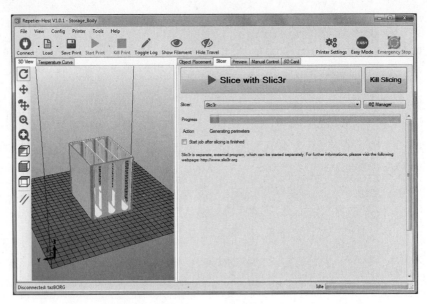

Slicing the model can take quite a while, so be patient!

Print the Storage Body

The next step is pretty predictable: you're going to print the part. But before you start the print, click on the **Preview** tab and take a moment to check the model to see if it looks like it's supposed to.

If it is right, go ahead and push that **Start Print** button! As I always do, I recommend you stick around to watch the first layer. If that one prints correctly, the part will probably turn out well.

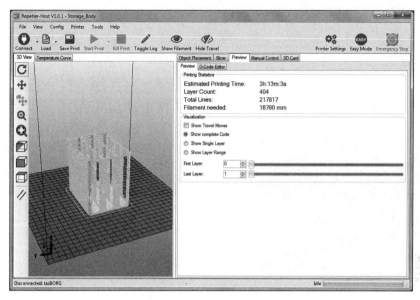

If it looks right, print it.

Load the Drawer .STL File and Slice It

After your first print has finished, you can pull it off the printer and get started on the drawers. Most likely (though it depends on your settings), your printer's hot end and heated bed turned off after the print finished. If they did, you need to preheat them again.

While they're heating up, you can load the storage drawer model by clicking **Load** and selecting the **storage_drawer.stl** file, which is included in the model pack at the Idiot's Guides website (idiotsguides.com/3dprinting). This one needs to be oriented with the bottom down, so it's pretty straightforward; any adjustments can be made using the tools in the **Object Placement** tab. This print will also use the exact same settings as the last one. Once you're ready to slice it, go to the **Slicer** tab and click **Slice with Slic3r**.

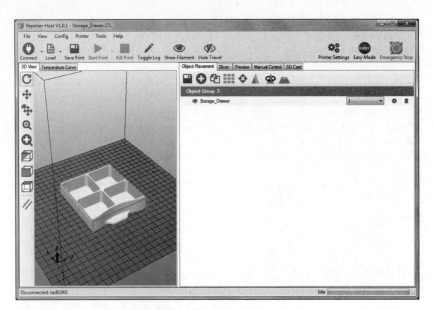

Orient the drawer with the bottom down.

HOT TIP

You may want to use a different color for the drawers for some lovely contrast. If you do, be sure to adjust the filament diameter settings to match the new filament.

Print the Drawer

If everything looks good after the slicing in the **Preview** tab, you can start the print by clicking **Start Print**. Just like with the storage body, you'll want to watch the first layer to make sure there are no obvious problems. You should then let it cool a few minutes before removing.

Project 4: Storage Box with Drawers

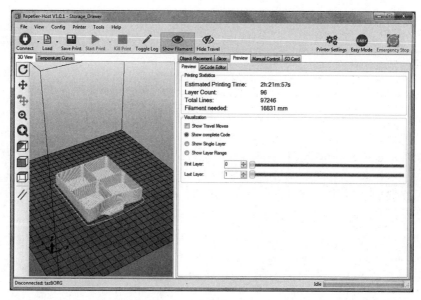

As always, make sure the preview looks right before the print, and watch the first layer after.

Print Two More Drawers

If the storage body and drawers turned out well, you'll probably want a couple more drawers to put in the other two sections of the storage body. To do this, you can just repeat the previous two steps to print two more drawers. Or, if you're feeling really adventurous, you can hop over to project 6 to learn how to design your own custom drawers! Either way, once each is done, allow to cool before removing.

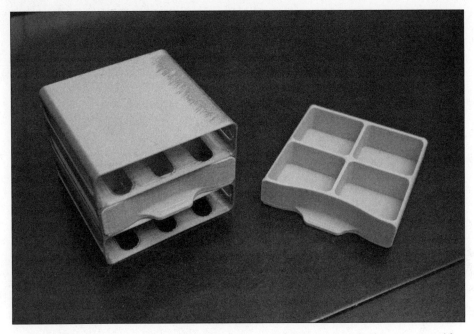

The printed parts. You can go to project 6 to learn how to make your own custom drawers to use with this body.

PART

4

3D Modeling

Once you've learned how to 3D print, you'll be eager to print all of the cool models you find online. But it won't be long before you want to design your own models to print. In this part, I teach you how to do just that. You learn how to use CAD software to create your own models, and how to optimize them for 3D printing. You also learn about basic reverse engineering techniques, so you can print useful parts that work with existing products.

CHAPTER

15

Introduction to CAD

Computer-aided design (CAD) is how virtually all new products are designed these days. CAD software is used to create 3D models of the parts of a product, which can be used to make every step of the development and production process more efficient. Over the past 30 years or so, CAD has literally revolutionized product design and development.

To really take full advantage of 3D printing, you'll want to learn 3D CAD modeling. It's not an easy skill to learn, but it's necessary if you want to create your own designs. And learning the basics shouldn't be too difficult. In this chapter, I take you through all you need to know about CAD for 3D printing.

In This Chapter

- How CAD differs from other kinds of 3D modeling
- Common CAD modeling and sketching commands
- Why units and scale are important in CAD

Why CAD Came About

Before the introduction of CAD, developing a new design was a very slow and error-prone process. If someone had an idea for a product, the design had to be drawn by hand. Just creating the drawing introduced the possibility of mistakes, and that's assuming the design would work. It wasn't until the parts were actually made that the designer could know for sure whether they would even fit together, much less work.

FASCINATING FACT

Most modern professional CAD systems have add-ons for running simulations. Engineers can use these simulations to test heat transfer, stress analysis, load capabilities, and so on. With the proper use of simulations, the development process in many industries has been made significantly more efficient.

Modern CAD systems change all of that and drastically improve the process. 3D models can be created and used to test how the parts fit together. Those models can even be used in physics simulations, which can tell the engineer if the individual parts will hold up to the stresses involved, and if the assembly as a whole will work.

Once the 3D models have been created, they can be used to quickly create accurate 2D technical drawings for manufacturing. Because those drawings are based on the models, there is little risk of inaccuracy. Creating those drawings is a lot faster than drawing them by hand, and they can be automatically updated if the 3D model is revised.

In many cases, like when the parts will be 3D printed or CNC milled, the drawing isn't even necessary. Instead, the 3D models can be sent straight to the 3D printer. This makes drawings unnecessary in the prototyping stage (although most companies still use drawings for manufacturing).

Basically, CAD has made every step of the engineering process better. But it's not just engineers and designers who can take advantage of the benefits of CAD. As a 3D printer user, you, too, can use CAD software to create designs to print.

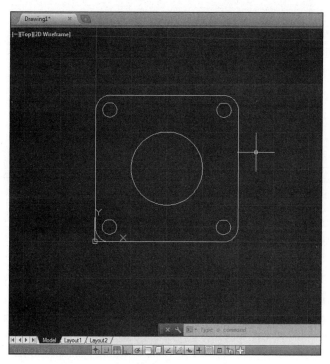

A simple flat pattern designed in 2D CAD.

Artistic 3D Modeling vs. CAD Software

Technically speaking, CAD is not synonymous with 3D modeling (though it is often improperly used that way). A great deal of CAD work is done purely in 2D, in which case it's similar to traditional *drafting* by hand (just using a computer to draw instead of a pencil). And conversely, not all 3D modeling is CAD.

> **Drafting** is the process of creating technical drawings of parts and assemblies. A person who does this is a drafter (or draftsman or draughtsman). Traditionally, this was done with a pencil and paper, but it is now done almost exclusively with CAD software.

CAD is a term usually used in the context of engineering. CAD software is generally used to design functional parts and assemblies. This is in contrast to 3D modeling software that's used to create artistic models like sculptures and characters.

So why the differentiation? It's not just the usage that makes them unique; the software itself is vastly different as well. Because artistic 3D modeling and CAD modeling serve completely separate purposes, the respective software is optimized for those purposes.

Artistic 3D modeling software, or sculpting software, is made to make it easier for artists to do their jobs. Exact dimensions are rarely important to them; instead, they need to be able to freely form the models visually. Their goal is to create a model which is aesthetic in nature, so their 3D modeling tools reflect that.

CAD modeling, on the other hand, is all about creating models with very specific dimensions and a defined form. CAD designers don't approach this from a visual perspective, but from a mathematical one. The aesthetics of the part being designed can certainly be important, but they still have to be specifically defined.

The differing goals between artistic and CAD 3D modeling result in a fundamental difference in the way the software works. Artistic 3D modeling software is free form and allows the artist to create and modify shapes, points, and faces without necessarily defining any parameters. The artist can just sculpt the model based on how he thinks it should look.

3D CAD software is parametric, which means that every shape is defined with specific parameters (or dimensions). Technically speaking, some shapes can be created without explicitly defining parameters, but doing so is generally considered to be bad form. And whether the user defines the parameters or not, they are still created by the software and can be modified later.

HOT TIP

If you'd prefer to use an artistic 3D modeling program instead of a CAD program, Blender is a good option. It's also free and open source, and has a huge user base and plenty of add-ons. It's a very high-quality program, so much so that it competes with the professional alternatives. However, it has a very steep learning curve and is difficult to learn.

The ability to modify existing parameters is probably the most important function of CAD software. Because every feature you create is defined by parameters, those parameters can be modified to change the model. Even if it's the very first feature you create, you can still change it, even after the rest of the part has been modeled.

From a design perspective, this trait of CAD software is invaluable. You can easily adjust the dimensions of the parts as needed during the design process. Parameters can also be linked together with math formulas, so that adjusting one dimension automatically updates others. Using that functionality, you can create a single model that can be used to produce parts in a variety of sizes (for example, nuts and bolts in different sizes).

CAD Software Options

When it comes to CAD software, the best programs you'll find are the ones made specifically for engineering. The three most popular parametric engineering CAD programs today are Solidworks, Autodesk Inventor, and Pro/Engineer (which has been renamed to Creo). These are all great programs, but because they're meant for professional use, they're very expensive. However, they often have student versions available, so that can be worth looking into if you're a student.

However, if you're not a student and don't want to spend a small fortune on software, there are still some options available. In recent years, a number of open-source and/or free CAD programs have been released, such as FreeCAD, Tinkercad, and Autodesk 123D. These are all usable, and free is a hard price to beat, but I've found them to be pretty lacking in comparison to the professional options. That said, they should be adequate for hobby use, especially when you're just getting started.

Aside from CAD applications, there are some other 3D modeling programs you may find useful. These won't be parametric like the others, but many people still find them useful. SketchUp, Blender, and Wings 3D are some good examples you may want to try if you find the CAD workflow doesn't work for you.

Another free (and open-source) option is OpenSCAD, which is very unique among CAD programs. The developers advertise it as "The Programmer's Solid 3D CAD Modeler," because you build the model by writing a script which defines the model instead of using a graphical user interface (GUI). It may sound strange, but it's proven to be very popular with people who are comfortable with code.

 HOT TIP

Try using a variety of different programs to get a feel for what each offers. Even the programs that aren't free will usually have free trials available. This software is difficult to learn, and it's a good idea to try different ones to determine which makes the most sense to you.

An Overview of Common CAD Program Commands

I don't have room here to cover all of the commands used in all of the various CAD programs, but I will cover some of the more common ones. Bear in mind that the command names will vary depending on the specific software you're using, but the basic functions of the commands will remain the same. The specific method for using the commands will also differ between programs, so I'll really just be explaining their functions here.

I'll break these up into two sets of commands: modeling commands and sketching commands. The modeling commands are used to add new features to the model, while the sketching commands are used to define 2D drawings to use with those features.

Modeling Commands

Extrude: This is almost definitely going to be the most common command you'll use. Its name is derived from the extrusion manufacturing process, which might help you understand what it does. (See Chapter 3 for a refresher if you don't.) When using this command, you sketch a 2D cross-section, which is then extruded (perpendicular to the cross-section) to form a 3D feature.

Extruded cut: In some CAD programs, this command is combined with the extrude command (you just specify that it's a cut). This works in the same way as the extrude command, except it removes material from an existing solid body (to create a hole, for example).

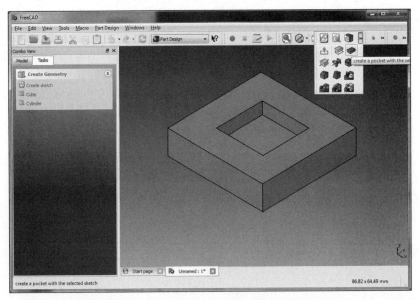

An example of the extruded cut command in a CAD program.

Revolve: This command allows you to create a feature by spinning a 2D cross-section around a central axis. You define how many degrees it will spin, and a solid body is created based on that. So, for example, to create a sphere, you could rotate a half-circle 360°.

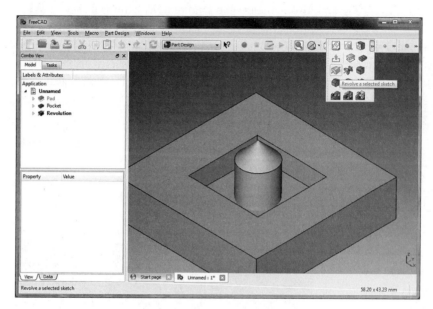

An example of the revolve command in a CAD program.

Revolved cut: Just like with the extruded cut, this command does the same thing as the revolve, except it removes material instead of adding material.

Sweep: A sweep is a little more complicated than the extrude and revolve commands. This creates a solid by taking a 2D cross-section and sweeping it along a 2D or 3D trajectory (the path). This requires a sketch for the cross-section, as well as another sketch for the trajectory. The sweep command is often used to create pipes and tubes but can be used for a variety of purposes.

Swept cut: You may be noticing a trend here. Most commands that can create a new solid also have a counterpart for removing material from an existing solid, so sweep has a counterpart in swept cut. This command has many uses, like cutting channels into a part.

Loft: This command is similar to the sweep command, except you sketch two or more cross-sections (instead of single cross-section and a path). This means the cross-sectional shape of the solid can vary along its length (for instance, from a circle to a square). If, for instance, you wanted to transition from a square tube to a round tube, you could use this command.

Lofted cut: As usual, the loft command has a material removal counterpart. If you're familiar with a Venturi (like in a carburetor), this is the command that could be used to create such a feature.

Hole: Many CAD programs also include some kind of hole command. You could really make the same feature with the other commands, but this often simplifies the process. Often, specifications for a number of standard fasteners will be automatically loaded, allowing you to quickly create holes for those fasteners. However, the fastener threads aren't usually physically modeled; instead, a visual representation (usually a graphic) is added along with the specifications. This means the threads made with this tool won't be 3D printed.

Fillet: Use this command to create rounded edges. It can be used on both interior and exterior edges. It is defined based on the radius of a circle, which is *tangent* to both faces that intersect to make the selected edge.

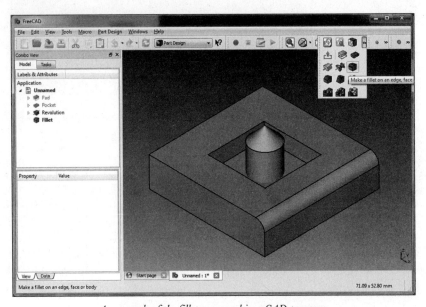

An example of the fillet command in a CAD program.

 **DEFINITION**

In geometry, a **tangent** line touches a curve at a single point and continues straight on from that point. The easiest way to visualize this is to picture a circle. A tangent line touching the far-right side of the circle will continue on vertically (up, down, or both), touching just a single point of the circle. The actual mathematics of defining a tangent line are fairly complicated, but luckily the CAD software handles all of that—you only need to be concerned with the practical effect.

Chamfer: This is similar to a fillet, except it cuts the edge instead of rounding it. It's useful for making edges less sharp without rounding them. It's defined based on the distance it's cut in on each face, or one distance and an angle (commonly 45°).

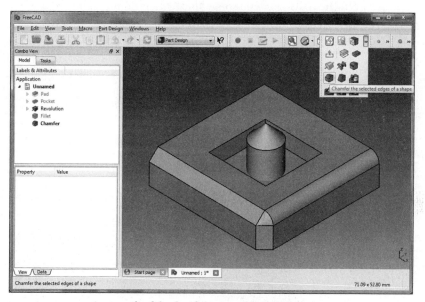

An example of the chamfer command in a CAD program.

Shell: You can use this command to make a solid body hollow. For example, you could use it to make a cube into a hollow box (with one side removed and the others given a thickness). The shell command requires that you select a face to remove and define a thickness to give to the other faces. This is very useful for creating parts like project enclosures.

Mirror: The function of this command is pretty straightforward: it allows you to take a feature and reflect it across a plane. You use this to create symmetrical features like bolt holes.

Helix/spiral: This really isn't a modeling command on its own but is more of an intermediary. The helix command can be used to create various kinds of spiral paths, which can in turn be used to create things like springs and threads.

Plane: This is a command for creating a reference plane in 3D space. It can be used as a mirroring plane, as a plane to extrude from, or simply as a dimension reference.

Axis: Like a plane, this is used for reference. It can be used for dimensioning, as an axis for the revolve command, or for creating other references (like planes).

Point: Another reference, this one is used to create a single point in 3D space. For instance, you could create a point at the intersection of a plane and an axis. It's useful for creating dimensions, and to use as a reference for other features.

Sketch: Most features will require a 2D sketch to define their cross-sections, and this command allows you to create that sketch. A sketch can be made on any plane; once you've selected a plane, you'll have a whole new set of commands that are used specifically for drawing.

Sketching Commands

Line: The most basic of all of the sketching commands is the humble straight line. It's pretty self-explanatory: you use it to make lines. Draw four lines in a square, and then extrude that, and you'll have a cube.

Circle: Again, this one is pretty obvious: you use it to make circles. They can be defined in a number of ways, such as by radius, diameter, center point, tangent lines, and so on.

Ellipse: As you'd expect, this is used to draw ellipses. You specify the height and width separately as needed.

Arc: Both circles and ellipses can be trimmed to make arcs, but you can skip that extra step and draw an arc from the get-go. Just like circles and ellipses, arcs can be defined in a number of different ways that are useful in different situations (such as by their end points and radii, three points, and tangencies).

Rectangle: If you don't want to bother drawing four lines manually, you can use this to quickly create squares and rectangles.

Polygon: To create a polygon (like a hexagon or octagon), you could manually draw it, but that would take a significant amount of time to properly define the angles and lengths of each line segment. So many CAD programs include this command to quickly create polygons with a specified number of sides.

Text: The text command is useful for engraving or embossing letters onto a face or for creating cut-out text.

Fillet: Just like you can round edges when working with the 3D model, you can also round corners when sketching. The command works in the same way by defining a radius.

Chamfer: You can also *chamfer* corners while sketching. Again, this works in the same way as when you're using it on the 3D model.

 **DEFINITION**

A **chamfer** is a simple beveled edge that connects two surfaces. If the surfaces meet at 90 degrees, a standard chamfer will cut across at 45 degrees for symmetry. However, a chamfer does not have to be symmetrical and can cut across at other angles as well.

Trim: This command is used to trim a line, circle, ellipse, and so on before or after an intersection (or between two intersections). This is an extremely useful tool, as it allows you to combine primitive overlapping shapes into complex cross-sections.

Extend: The opposite of the trim command. This extends a line along its current path until it intersects another line.

Offset: The offset command is used to create another line that is offset from an existing line. For example, to create a hollow tube, you could draw a circle and offset it by the thickness of the tube wall. When you extrude that, it will create your hollow tube.

Mirror: This is for mirroring lines across a mirror line. It's very similar to the mirror feature used in the 3D modeling area.

Pattern: How exactly the pattern features work depends on the CAD software you're using, but they all have the same goal. Its job is to take a part of your sketch and create copies of it in a specified pattern. This is useful for creating things like vent holes.

Dimension: Dimensioning your sketches is extremely important. If your sketch isn't completely defined, it can easily be accidentally distorted. It's also essential that you add dimensions everywhere on your sketch to make sure it's the correct size. How can you create a true cube if you don't ensure that every side has equal dimensions?

Constraints: Constraints serve a similar purpose to dimensions; they're used to keep a sketch defined exactly as you intend. They can be used to keep lines parallel, perpendicular, equal in length, and so on. Some CAD programs attempt to automatically create constraints, while others require that you add them all manually. Either way, it's important to make sure they're there when necessary.

The Importance of Units and Scale

Before you start modeling anything using the commands I've just taught you, it's essential that you understand the concepts of units and scale. Both of these are basic factors in CAD design, which makes them critical to the proper use of your CAD software.

Choosing Units

The first thing you should do when you create a new 3D model is choose your units so the CAD program knows what you want. Do you want to work in inches or millimeters? Pounds or kilograms? For example, when you type in "1.25," does that mean 1.25mm, 1.25cm, 1¼ inches, or 1 foot 3 inches? It's not just a question of whether you want to use the metric system or not; rather, it's about the specific units you'll be using for dimensions and for other information like volume and mass.

The units you use for length are the most important thing to specify by far. That's because everything you dimension while you're modeling will be in lengths and angles; while the angles are universal, the lengths are not.

However, you can also change the units for volume, weight, and even time. Volume and weight are useful for determining things like the weight and volume of the part, density of materials, and so on. Time is really only used for simulations, though, so you shouldn't worry too much about it.

> **FASCINATING FACT**
>
> The proper use of units will allow you to do a lot of useful things. For example, you can determine the weight of the part you design, its volume, and its center of gravity. This can help you figure out how much filament you'll need and how the part will perform.

So what units should you use? If you're American, you're probably the most comfortable with inches. However, there is a strong case to be made for using millimeters instead. First of all, the math is just a lot easier when you use the metric system. Fractions of an inch can get pretty cumbersome, and 12 inches to a foot doesn't make for easy math.

But it's not just the math that makes the metric system better for CAD modeling. When you export the model into the .STL format for 3D printing, the units you modeled it in aren't saved. So they have to be specified by the slicing software when you import the .STL. Most slicing software uses millimeters by default, and most people export their models in millimeters as well. The .STL file doesn't store what kind of unit was used in its creation, so it's necessary to make sure the exported unit type matches the imported unit type.

While the model can be converted to millimeters when you export it, that's just adding another step. Personally, I think it's best to use the metric system from the beginning so you're comfortable with it. However, if you really want to stick with using inches, you certainly can.

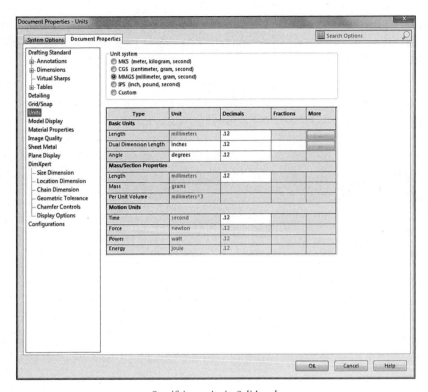

Specifying units in Solidworks.

Scaling in CAD

Once upon a time, when engineering drawings were still created with a pencil and paper by drafters, scale was a practical concern. If you wanted to draw a building on a piece of paper that would fit in your briefcase, you obviously couldn't draw it at its actual size. That's where scale came in.

In order to draw something large at a practical size, the drafter would draw it at a fixed scale. For a building, he might draw it at a scale of "⅛ inch = 1 foot 0 inches," meaning that every ⅛ inch on the paper represented 1 foot on the actual building. So a wall 10 feet long would be drawn as 1¼ inches long on paper. The drafter had to take this into account while creating the drawing in order to keep the entire drawing at the correct scale.

However, with the introduction of CAD, this was no longer necessary. In 2D CAD, the drafter could draw the building at a full 1:1 scale and simply rescale it as needed for printing. CAD software could handle rescaling the drawing automatically, so the drafter would only ever have to draw things at a 1:1 scale and let the software do the rest.

But some habits die hard. For people who learned drafting with a pencil and paper, the idea that things had to be scaled tended to stick. So many people continued to carry on with that habit and still drew things at different scales even in CAD programs where it wasn't necessary to do so. That continues even to this day, despite the fact that it's entirely unnecessary (and even causes confusion).

Even more bizarrely, that habit has even managed to make its way into 3D modeling in some cases. It's not common, but some people still make the mistake of defining the dimensions at a scale other than 1:1. It's important that you learn from the very beginning not to do this, because it'll cause a lot of headaches if you do.

When you're using CAD software, always create dimensions at a 1:1 scale. If something is supposed to be 100mm long, give it a dimension of 100mm. This shouldn't be hard to do, because it's the natural thing to do anyway if you're new to CAD and drafting.

If for some reason you need to create a drawing and the model doesn't fit on the page at 1:1 (or it's too small), scale the drawing itself instead of the model. If you want to 3D print a miniature version of something you modeled, you can scale it in slicing software (unless it's intended to always be miniature). Basically, the best way to approach this is to just not think about it and dimension everything at its actual size.

The Least You Need to Know

- CAD software is designed for creating models with specific dimensions, while artistic 3D modeling software is useful for sculpting free-form models.
- Specific CAD commands depend on the software you're using, but most have similar sets of commands.
- Units are important, and my recommendation is to use the metric system, because it's what the 3D printing community tends to use.
- Always create all CAD dimensions at a 1:1 scale. If any scaling needs to be done, it should be done separately in the drawing or the slicing software.

CHAPTER

16

Modeling Techniques and Best Practices

Just knowing the CAD commands isn't enough if you want to have an enjoyable 3D modeling experience. It's also important to get a good grasp of the workflow used by CAD designers. With a good workflow, you'll be able to design anything with enough patience. Everyone has their own modeling style, and you'll develop your own. But with this chapter, I give you a good starting point to get you on your way.

In This Chapter

- Designing assemblies
- Tips for successful 3D modeling
- How to create models for 3D printing

Premodeling

Before you actually start modeling anything, you want to have a good picture in your mind of what it is you want to model. You can't just start modeling if you don't have a clear idea of what the finished object should look like. (Well, technically I suppose you could, but it would be a frustrating experience.) In fact, it's not uncommon to spend more time thinking about what you're going to model than actually modeling it.

Take your time with this process, and don't just jump right into the modeling. If you start before you're ready, you could easily spend more time fixing the model than if you had just been patient and done it right from the beginning.

Once you've come up with a clear idea of what you're going to model, it's time to open your CAD program and get started. There are a million ways to model the exact same part, and how you go about modeling it is part of your style. Everyone will take different approaches to the same model, and there isn't necessarily any right way.

That said, there are some basic techniques which most people use. The idea is to start with large features to create a rough shape, and then add features to refine it. As you're modeling those features, be sure to add dimensions and constraints to fully define the sketches you create.

In order to add those dimensions, you should already have an idea of what they should be before you start modeling. That's part of the premodeling thought process and helps to speed up the actual modeling.

Tricks of the Trade

When I'm not writing delightful books on 3D printing for you to read, I work as a mechanical designer and drafter. This means my job is to do 3D CAD modeling and to create technical drawings. In my time doing this, I've picked up a few tricks to make the modeling process more efficient, and I'm going to teach them to you!

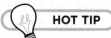

To determine if a particular feature is essential, ask yourself if the part could still function without it. In most cases, a particular part will only have a handful of essential features. The rest are really only there to improve the aesthetic and feel of the part.

Separate the essential features from those that are mostly just cosmetic. Start with getting the functionality of the model right before you start in with those features that are really just there to make the part look finished. Things like fillets and chamfers are usually just added to make edges less sharp, so save those for last. On the other hand, things like bolt holes and mating surfaces are essential to the functionality of the part, so concentrate on getting those right first.

If you can, try to prioritize the features based on how "hard" they are. By that, I mean features that have strict requirements should come first. If, for example, you're designing a replacement bracket for a shelf, you'll want to model the screw holes as soon as possible to make sure they're right. After that's done, you can move on to adding things like supports, and then finishing touches like fillets.

Start with important features, like mounting holes.

Get comfortable with the commands that are designed to speed up the modeling process. If your part is going to be symmetrical, there is no sense in modeling the entire thing. Instead, just model one half and then mirror the features to the other side. That has the added benefit of ensuring that both sides are exactly the same and reduces the chances of making a simple mistake (like entering the wrong dimension).

The pattern tools included in most CAD programs can also be very helpful in this regard. If identical features (like holes or slots) will be used in regular intervals, that's the perfect situation to take advantage of the pattern tool. This isn't just for linear or rectangular patterns either; most CAD software has the ability to create circular patterns, which is great for modeling things like gear teeth.

In order to take advantage of all of these tools, you should take the time to explore your CAD software every now and then. Most CAD designers tend to get comfortable with certain tools and will often use those even when another tool would work better. So don't be afraid to experiment with different commands, so you'll be familiar with them when the right situation arises.

Keep things simple. It's better to use two simple features than one complex feature. Doing so will make errors less likely and will make the features easier to modify later if needed. If you ever share the model, it'll also make it easier for other people to follow your work.

If you have any experience with programming, you'll probably recognize the importance of this. Complex formulas and statements may take up less space, but they're also difficult to decipher and modify. The features in a 3D model can be similar. You could draw a really complex sketch to do as much as possible in a single feature, but there really isn't any benefit in doing that. Instead, you could just break it up into a few simple sketches that will be easy to understand and modify later.

You'll probably need to modify your model. Relating to my "keep things simple" tip, it's rare to get the model exactly right on your first try. The chances are good that you'll need to change some dimensions at the very least, and possibly remodel some things completely.

To be able to make those modifications without a lot of trouble, you'll want to use good technique while modeling. Set up all of your constraints properly and define all of your dimensions. Don't get sloppy, because you'll just be making it harder on yourself when you have to modify the model later. CAD modeling isn't a race; take your time to do it right, and you'll save yourself some trouble later.

Sometimes it's best to just start over from scratch. This tip may be the hardest to swallow. If you've spent a lot of time modeling something, it can be difficult to just give up on it, but that's often the best choice. As I've mentioned, there are a million ways to model something. You may not realize until you're finished that you chose the wrong way to model a particular part; it happens to all of us.

If you modeled the part in a way that is less than ideal, it can be difficult to make adjustments. Trying to make adjustments to a bad model will often take more time than simply remodeling the entire thing. One of the best skills you can learn is how to know when to just start over with a clean slate.

Assemblies and Fitting Parts

Modeling a single standalone part is usually a relatively simple matter. But things get significantly more complicated when you design assemblies. An assembly is simply a group of parts that are designed to fit together. For example, a car engine is an assembly. And, technically speaking, the car itself is an assembly, too.

But the terms aren't really important. What's important is the idea that assemblies are systems of parts that are designed to fit together.

As an example, picture a regular old pair of scissors. Think of the all-metal kind, not the kind with plastic handles. A pair of scissors like that is a very simple three-part assembly, made up of one blade and handle, the second blade and handle, and the rivet holding them together. But even a simple assembly like that takes some planning.

Each blade and handle has to be designed as you'd expect: to have a sharpened edge and comfortable handle. But it also has to be designed to fit with the other piece to make a functional pair of scissors. That means the blades have to be designed to line up properly, at the same time that the two handles touch.

As you add more parts to an assembly, things get exponentially more complicated. A mechanical wrist watch, for example, is one of the most complex mechanical assemblies ever devised. There are a bunch of parts that all have to fit together perfectly in a very small space. Designing something like that is mind-bogglingly difficult, because there are just so many interfacing parts to plan.

FASCINATING FACT

As complicated as mechanical wrist watches are, they're even more impressive when you consider the times in which they were designed. The parts and assemblies were all sketched with a pencil and paper, and the parts were mostly made with basic hand tools. Even with our modern software and tools, replicating a mechanism that complicated is very difficult.

As you can see, designing assemblies is no easy task; I'm not talking about just putting Legos together. Each part has to be designed to perform its function, and also to connect with the other parts in the assembly.

Planning an Assembly

So how should you go about designing an assembly with multiple parts that need to fit together? The answer is careful planning, and unfortunately there is no quick and easy trick to this. This is what mechanical engineers and designers are paid to figure out, so don't let it bother you if you find the process difficult.

This planning stage is one area where 2D CAD programs can come in handy, even if the actual parts will be modeled in 3D. You can draw out a representation of the parts in the assembly to get an idea of the dimensions of the individual parts.

To illustrate this, imagine you had a 100mm-long belt and two pulleys with diameters of 10mm. You want to determine how far apart the pulleys need to be mounted in order to keep the belt taut.

Trying to figure this out while modeling the parts would be fairly difficult, but you can do it quickly and easily in 2D CAD. You can draw one pulley as a circle, with another half-circle around it to represent that part of the belt. You can then measure the length of that half-circle (in this case, 16.4934mm). We know that there are two pulleys, so you multiply that by 2 and get 32.9868mm. Subtract that from 100mm, and you get 67.0132mm, which is the rest of the belt. Divide that by 2, and you get 33.5066mm.

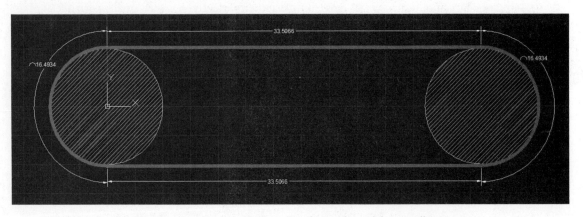

Sketching in 2D to come up with dimensions for modeling a belt and pulley system.

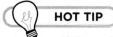

HOT TIP

If the pulley example is a little difficult to wrap your head around, try thinking about why it works. The belt is wrapping around two pulleys, so it will be in contact with half of each one. So you take half of the perimeter of each pulley and add that. You know the belt is 100mm, so you subtract the perimeters from that. You then divide the result by 2 in order to get the distance between the pulleys (since the belt is a loop). Drawing this just saves you from having to do the math manually.

So you now know the pulleys need to be mounted about 33.5mm apart in order to keep that 100mm-long belt taut. With that information, you can design your assembly knowing exactly where your pulleys need to be mounted. This removes the guesswork from the process, which is key to successfully designing assemblies.

This illustration, of course, is a very simple example of the kind of thing you'll need to plan for. In the real world, assemblies are usually much more complicated. That's why you'll need to take the time to carefully plan out the assembly before you start modeling it.

Fitting Parts Together

But what about just designing two parts that need to fit together? Anytime you design an assembly, you'll have to think about how you're actually going to connect the various parts to each other. Are you just going to bolt them together? Do you want to design some kind of clip system? What about parts that need to fit together like puzzle pieces? You can generally break this down into two basic types: parts that will be fastened together and parts that need to fit together.

Fastening parts together is actually a surprisingly complicated topic. Take a stroll through the fastener aisle at your local hardware store and just look at how many different kinds of screws, bolts, nuts, clips, nails, and so on there are. The reason there are so many types of fasteners out there is because there are all kinds of different requirements when it comes to putting parts together.

But for our purposes it's best to keep it as simple as possible. Here are some different ways to fit parts together:

- If you're just attaching two parts together and there won't be a lot of stress, you can just use undersized holes and screw some machine screws into them. Because the plastic is relatively soft, it will basically just cut its own threads.

- If your assembly requires a bit more strength, you can use heat set inserts. These are little metal tubes with threads on the inside and ridges on the outside. You heat them up and push them into a hole in the plastic. As you push it in, it will melt the plastic around the insert, and then it will cool to lock the insert in. You'll then have some nice metal threads for your screws.

- It's possible to avoid using screws altogether in some situations. The battery cover on the back of your TV's remote control is a good example of this. Instead of screwing the battery cover on to the body of the remote, a small, springy tab is used to hold it in place. You can design similar tabs to hold your 3D-printed parts together.

> **FASCINATING FACT**
>
> There are many such techniques like the springy tab that can be used to connect parts together without having to use fasteners. Woodworking is full of these ingenious joinery methods that don't use fasteners, so that may be one area you can look into for inspiration. You can also just look at the products around your home for ideas. You may be surprised by how many techniques have been developed over the years.

Fitting parts together without fasteners is a whole other skill. It's one thing to simply bolt two pieces of wood together, while it's something else entirely to join them together with something like a tongue and groove. When it comes to 3D printing, how parts actually fit together can be a little frustrating. Say you wanted one part to fit inside other, like a piston in a cylinder. To get a tight fit, the piston has to be just slightly smaller than the inner diameter of the cylinder. The question is: how much smaller?

That question is a major one in the engineering world, and it really depends on what you're trying to accomplish. If the parts will be moving (like a piston), it can't be too tight or the friction can be a problem. On the other hand, if the parts aren't supposed to move (like a dowel pin), it needs to be very tight.

But 3D printing throws in another curveball. Because the tolerances involved with FFF 3D printing are so loose (easily +/-0.1mm, but often much more), it's hard to get a precise fit. So you'll need to do some experimenting to determine how much of a gap to leave between parts for different applications. This will depend on your printer, the type of fit you're trying to get, and even on the particular part you're printing.

Modeling Successful 3D Parts

When you're modeling parts for 3D printing, there are a few things you should do to ensure they can be printed successfully:

Avoid supports, if possible. You can really design the part however you want, if you plan on using supports. But ideally, you'd want to design the part so you can print it without supports. Printing without supports generally yields higher-quality parts.

Make sure the part isn't too big for your printer. That includes anything that's added in the printing process, like rafts, brims, or supports. Those can easily add 10mm to all sides of the part, so take that into account when designing the part. If it's a tight squeeze, you may not be able to use brims or rafts, which could affect how well the part prints.

Take steps to limit warping. Warping can be a very difficult problem to overcome, and it gets worse as the part gets larger. That's because warping is caused by the contraction of the plastic as it cools. The longer and taller the part is, the more material there is contracting and the more warping there is.

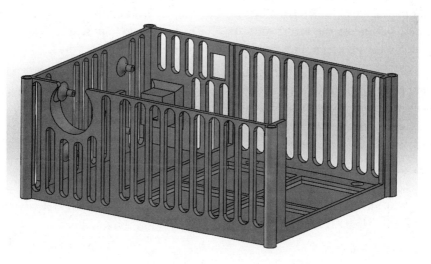

Slots were added to this electronics enclosure for ventilation and to limit warping.

There is something you can do to limit that warping, though. Because warping is caused by the contraction in long pieces of plastic, you can fight the warping by interrupting those long, solid pieces of plastic with holes or slots. The idea is just to break up any long, solid stretches, so the contraction only occurs in short areas to reduce the effect.

Of course, that's not always possible. If that's the case, you may have to use a material that is less prone to warping (in other words, materials that don't contract as much). PLA is a good option, because it's significantly less susceptible to warping than ABS.

> **HOT TIP**
>
> These challenges can sometimes be difficult to overcome, but it shouldn't take long for it to become second nature. Compared to other manufacturing processes, 3D printing actually has very few design constraints. Just keep in mind how the design of your part will be printed.

Exporting Files

Once you've finished modeling your part, you'll need to export it for 3D printing. As you've learned in previous chapters, you need a .STL file for the slicing software to work with. Luckily, all CAD software can easily export models into the .STL format.

When you export the model, you will then need to specify a few things for the .STL file. The first thing you'll need to set is the units. I've mentioned previously that .STL files do not store unit information, so that has to be set when you export the file from the CAD program and when you import it into the slicing software. The exported and imported units have to match. Most slicing software uses millimeters by default, so you should export the file in millimeters (unless you have a reason not to).

You'll also need to specify the quality you want in the .STL file. This essentially determines how many triangles will be used to make the surfaces of the models. More triangles mean higher-quality surfaces. This is especially apparent with models that have curved surfaces.

Chapter 16: Modeling Techniques and Best Practices

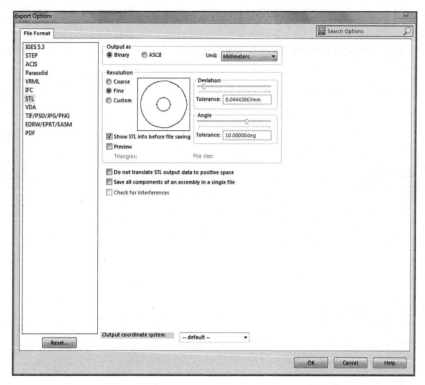

The .STL export settings in Solidworks, including the options to set the resolution (for the quality of the model).

If a model with curved surfaces is exported with a low number of triangles (resulting in a small file size), those curved surfaces could come out looking faceted. Instead of a nice, graceful curve, you'll have a series of flat segments making up the curve.

In order to make sure your printed parts come out at the highest quality possible, I recommend you turn the quality settings all the way up when you export the .STL file. This might create an .STL file that is very large. So, if you're planning on sharing the file online, you may need to turn the quality down. Otherwise, there is no real reason not to use high-quality settings on a modern computer.

The Least You Need to Know

- Always plan your model out ahead of time, especially if you're modeling an assembly.
- Sometimes it's better to just start over than to try and salvage a bad model.
- 3D printing has some unique considerations when it comes to modeling. This is especially true when it comes to reducing warping, which happens when the material contracts as it cools.

CHAPTER 17

Practical Reverse Engineering

In the context of 3D printing, reverse engineering is the practice of duplicating the design of a part or assembly. It's easy to see why this is a useful skill to learn for the 3D printing enthusiast. If you want to reproduce a part or interface with an existing part, you'll need to know how to reverse engineer it.

Luckily, reverse engineering simple parts probably isn't quite as difficult as you're expecting. It's a relatively simple process of taking measurements and 3D modeling the part based on that. However, actually taking accurate measurements can be a little tricky in some cases, so I've given you some helpful tips and hints in this chapter to help you out.

In This Chapter

- Why you'll want to learn reverse engineering
- Measurement tools that are useful and affordable
- Techniques for determining part dimensions
- The two ways you can model a part

Why You Should Learn Basic Reverse Engineering

One use for 3D printing that many people are interested in is the ability to replace broken parts they come across. Broken shelf brackets, car parts, and really anything made of plastic can be replaced with a 3D-printed part. Spending 50¢ on filament is sure better than buying overpriced replacement parts from the manufacturer (if they're even available).

In some cases, you may be able to find some of those parts online that someone else modeled. But most likely, the majority of the things you'll want to replace haven't been modeled yet. For that reason, you'll need to learn how to do it yourself.

Reverse engineering a part may sound a bit overwhelming, but it's not as bad as it sounds. All you'll be doing is 3D modeling a part as closely to the original as possible. You'll just be taking measurements and making sure the important measurements are as accurate as possible.

> **Reverse engineering,** in a general sense, is the process of determining the function and/or design of a man-made object or system. This can be anything from reverse engineering software to complicated mechanical systems.

So which measurements are the important ones? They're the ones for things like screw and bolt holes, the overall shape and size of the part, and any functional features. For most things you'll probably want to replace, there will probably only be a few features that have to be exact. This means there is almost always some room for inaccuracy on the unimportant features.

Finding the Right Measurement Tools

There are a wide range of tools used to measure parts for reverse engineering. Professional engineers use some expensive tools like comparators, go/no go gauges, or thread gauges. These are precision tools (with the gauges needing to be bought in sets) and are usually quite expensive. In a professional environment, they also need to be regularly checked to ensure that they are within spec. Luckily for hobbyists, though, most of those aren't necessary for the kind of reverse engineering you'd be doing at home.

The following describe the two most commonly used measurement tools by hobbyists; as you'll learn, one is a necessity for measuring, while the other is one you should probably stay away from.

Digital Calipers: A Necessary Tool

Calipers are a simple measurement tool that measure the distance between the arms of the slide. Even in the professional world, digital calipers are what engineers use most of the time. They're a surprisingly versatile instrument. Digital calipers can be used to measure interior distances, exterior distances, and usually depths as well. With simple distance measurements, most other measurements (like angles and radii) can be deduced.

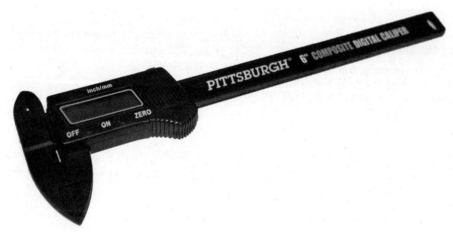

A very inexpensive pair of digital calipers.

These days, simple digital calipers can be purchased for less than $20. Even at that price, they're still reasonably accurate. The accuracy of digital calipers is usually within 0.02mm, which is better than most 3D printers are capable of anyway.

> Don't bother spending a lot of money on your first pair of digital calipers. High-end digital calipers can be pretty pricey and are unnecessary for hobby tasks. Inexpensive calipers will do just fine for most people outside of a professional engineering environment.

3D Scanners: Not All They're Cracked Up to Be

If you've been investigating 3D printers, you may have also come across 3D scanners. 3D scanners are devices that optically measure thousands of points on the surface of an object. With those points measured, a computer can reconstruct the surface of a 3D object.

The idea is that you can just throw a part on the 3D scanner, scan it, and then print a perfect copy. The appeal of this is pretty obvious; it takes away all of the trouble of having to reverse engineer a part. But do 3D scanners fulfill their purpose?

The short answer, unfortunately, is no. This is for two big reasons: the accuracy of the scan and the usability of the resulting model. Both of these are a pretty big deal and are keeping 3D scanners from being particularly useful in engineering applications.

The accuracy issue is one which you can probably see the problem with pretty easily. Even high-quality scanners tend to result in imperfect models. That's especially true for complicated parts or parts with any internal geometry.

Because 3D scanners work optically, they can only scan what they can see. If there is any internal geometry, the scanner wouldn't be able to see it. So that geometry wouldn't be replicated and would be lost.

> **FASCINATING FACT**
>
> Tinkerers, hackers, and makers have been experimenting with the development of inexpensive 3D scanners for a while now. Prototypes have been made from the Microsoft Kinect and even as apps on cell phones. However, none of these are at the point yet where they're very useful for reverse engineering.

That leads us to the bigger issue: the created geometry can't be easily modified. The scan just produces what is essentially a surface mesh. No actual features are created, which means there are no features for you to modify.

If you want to modify the model generated by the 3D scanner, you'll have to modify the mesh. The same is true if you just want to fix the model, like adding internal geometry or fixing errors. To modify the mesh, you'll have to use those artistic 3D modeling programs I talked about before.

Measuring the Part

For some parts, measuring will be a simple process. For others, it can be significantly more complicated.

Let's start with a very simple part, like a cube 10mm to a side. To measure that part, you just close your calipers and set them to zero. You then just measure the distance between three sets of perpendicular faces. If it was a perfect cube, you'd get exactly 10mm between each set of faces. And, of course, to model that part you'd just make a cube that is 10mm in each direction.

Make sure to set your calipers to zero before measuring.

However, as you might imagine, most parts are much more complicated than a cube. To measure real parts, you'll have to measure the distance between a lot more points. But the basic principle remains the same: gather measurements and reproduce them in your model.

Doing this will take a decent understanding of geometry. In most cases, you probably won't physically be able to take measurements of every feature on the part. Some won't be reachable, while others will have curves or angles with can't be measured. That brings me to the most important reverse engineering skill you need to learn: how make inferences.

Inferences

When calipers are your only measurement tool, you've got to learn how to make some inferences about the part. The inferences you'll need to make can be broken down into roughly three types: geometric, design intention, and proportions. For most parts, you'll need to use all three to really get the part modeled right.

Geometric

The easiest kind of inference you can make is geometric. If you can remember your high school math classes, these are the types of problems where you're given two variables and have to find the third. They're based on understanding the mathematics of geometry.

Lengths and distances can be easily measured with the digital calipers, so you'll usually be trying to use geometric inferences to figure out angles and radii. Doing that might seem overwhelming at first, but remember that your CAD software can help you do most of the actual math. You can just draw lines with the information you know, and then let the software fill in the rest.

An example of making a geometric inference about an angle is a basic right triangle. You don't have to know the angle of the *hypotenuse*; you just have to know the length of two sides (since you know that one angle is 90°). In the case of the following image, the two sides were 10, and the software filled in the hypotenuse of 14.1421. If you use the Pythagorean theorem—where $a^2 + b^2 = c^2$ (with c being the hypotenuse)—to check this, you'll find this is correct. Similar inferences can be made about isosceles triangles, which have two sides of equal length, if you know the lengths of all three sides or two sides and the height.

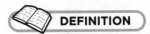
DEFINITION

The **hypotenuse** of a right-angled triangle is the longest side of the triangle. It'll always be the side whose endpoint doesn't touch the right angle.

Of course, you won't just be measuring triangles. However, the good news is that most features with angles can be broken down into simple shapes like triangles.

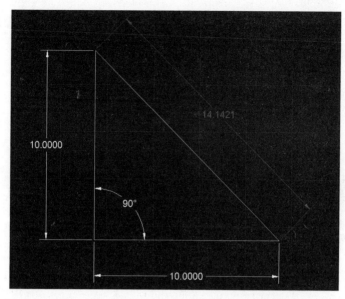

Two sides of this right triangle were entered manually, and the hypotenuse was then determined by the CAD software.

Design Intention

In my humble opinion, design intent is the single most important clue you have when it comes to reverse engineering. The principle behind this is simple: every part was designed by another person. People tend to design parts in fairly predictable ways, so you can make assumptions about the part based on that.

Remember the previous 10mm cube example? Let's imagine that instead of measuring exactly 10mm for each pair of faces, you actually measured 9.89mm, 10.01mm, and 10.8mm. You could model the part to those exact dimensions. Or you could ask yourself about the designer's intentions when he created the part.

Is there any reason why he would have given it those exact dimensions? If not, is it more likely that it was intended to be 10×10×10mm? Maybe the differences are actually a simple result of poor tolerances in manufacturing.

A case of figuring design intention. The measurement was 11.9mm, but we can pretty safely assume it was intended to be 12mm.

Unfortunately, it's not quite that simple. This time, let's say the measurements were 9.52mm, 9.53mm, and 9.51mm. You might chalk that up to poor manufacturing. But in reality, the designer might have been working in inches. ⅜ inch equals 9.525mm, so the designer might have intended it to be exactly ⅜ inch.

I bet this is starting to sound like a pretty big headache, right? How can you possibly figure this out when there are so many factors? One of the easiest ways is to first determine the origin of the part.

If it was designed anywhere but in the United States, it was probably designed in millimeters. If it was designed in the United States, it really could go either way. Some industries, especially older ones like the construction and automotive industries, tend to stick to using inches. Some newer industries, like the tech industry, often use millimeters instead. This may not help you determine for sure what units were used to design the part, but it should help.

Linear measurements aren't the only thing that can be determined based on design intent, though. You can use take that into account for every feature on a part, including curves. Does an angle look like it's about 45°? There is a very good chance it is. If a radius seems to measure out to about 3mm, you can pretty safely assume that was how it was designed.

The whole idea behind this technique is that you try and think like the original designer of the part. If you were designing that part, would you choose some odd number? Probably not; you'd most likely choose a nice, round number unless you had no other choice. You can use that fact to your advantage when trying to determine how you should model the part.

Proportions

And finally, you can make proportional inferences. These are judgments you can make just by eyeballing the part. To go back to the 10mm cube again, would you really need to take all of the measurements to know it is a cube? You could probably just take one 10mm measurement, and then notice it was a cube and derive the other measurements from that.

As usual, though, most parts aren't cubes and won't be this simple. Instead, you'll be looking at the proportions of the part to help determine the measurements. If there is a rectangular shape and you know that one side is 10mm and the other side looks to be about twice as long, you can pretty safely assume it's about 20mm.

Even if you can only determine one measurement, it's easy to reasonably infer the other dimension.

That's because as humans (you are human, right?), we're pretty terrible at estimating lengths and distances. But we're surprisingly good at making comparisons, like the proportions between features. I'm not sure exactly why this is, but it's true.

Whatever the reason, it's true that humans are much better at judging proportions than they are at judging actual measurements. So use that to your advantage when modeling parts. Don't try to guess all of the measurements; just guess proportions based on known measurements.

But why would you even need to be guessing at all? If you have those nice digital calipers, why even bother trying to guess proportions? The most common reason is when you're trying to model a part that you don't physically have access to, like if you're reproducing a part from a photo. You might be able to deduce some of the measurements from other objects in the photo, or you might be able to find some measurements online that other people have taken. If that's the case, you can use the proportions of the part to determine the missing measurements.

You can also use this technique for figuring out dimensions in areas you can't measure. This is especially true for interior features that just aren't accessible. Unless you want to try and cut the part in half, you'll need to make some inferences to get those measurements.

Modeling the Part

You've learned all of these methods for determining the dimensions of a part you want to replicate, so now it's time to model it. There are two ways you can approach this: model it as you measure it, or take all of your measurements ahead of time and then model it. Which approach you take is mostly just a matter of personal preference.

Personally, I tend to model the part at the same time I take the measurements. The main reason is that it's easier than having to try and write the measurements down in a sensible way. I think it also saves a little time in the process.

But writing down the dimensions can also work well. That way you don't have to keep switching back and forth between modeling and measuring. You can just take all of your measurements and then switch to modeling.

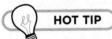

HOT TIP

When writing down dimensions, you'll always want some method for keeping them in order. You can do this by sketching a quick drawing of the part and writing in the dimensions. Or if drawing isn't your cup of tea, you can assign descriptions to the dimensions. But do something to keep them organized, so you don't accidentally get them mixed up.

Modeling a part when you only have digital calipers for measurements can be a little tricky. As I went over in the previous sections, there are often a lot of measurements that can't be taken with calipers. So this is when you'll want to start using the techniques you just learned.

Start by doing as much as you can with the measurements you were able to take. These will be your "hard" dimensions, which you can be reasonably sure about. Make inferences about the design intent here, like using round numbers unless it makes sense not to.

Once you've got all of those hard dimensions in, you can start filling in the blanks. In many cases, the geometry itself will give you the answers. Just like finding the hypotenuse of the triangles, a lot of the dimensions will fill themselves in based on the dimensions you do have. Or more accurately, those dimensions won't be needed to fully define the model.

Lastly, you'll need to make some judgment calls to fill in the dimensions you couldn't measure and can't be determined with simple geometry. These will usually be things like angles and radii for filleted edges. Luckily, these are the features that lend themselves the most to making inferences.

That's for a couple of reasons: they usually don't have to be perfect for the part to function properly, and they're fairly easy to get right without taking measurements. Fillets are often just cosmetic and aren't even integral to the function of the part. If they are important, it's okay because they're still pretty easy to guess at.

Angles are more likely to be important to the part's functionality. But they're also easier to determine. Designers usually stick to 15° increments unless they have a reason not to. You can usually correctly guess the angle just by looking at it.

In the cases where fillets and angles aren't obvious, your best bet will be trial and error. Experiment with different angles and radii until it looks like the real part. This is where the human knack for comparisons will come in handy. You may be surprised by how close you'll be able to get the part just by doing this.

Using these methods, you should be able to successfully reverse engineer most of the common plastic parts you come across. You'll be able to replace broken things around the house, on your car, or anywhere else something plastic breaks. The parts may not come out perfectly the first time you print them, but with a little refinement and practice, you should be able to produce functional replacement parts.

The Least You Need to Know

- Reverse engineering is a very useful skill for reproducing and replacing broken parts.
- Digital calipers are the only tool the hobbyist really needs to take measurements in most cases.
- For dimensions that can't be determined with digital calipers, you can make inferences and use deductive reasoning.
- When reverse engineering, always consider the fact that the part was designed by another person. Think about what that person would have done while designing it.

PROJECT 5

Monogrammed Coaster

Project Time: 20 minutes

For your first 3D CAD modeling project, you're going to start with something simple that you can proudly display on your coffee table (until your spouse makes you move it). You're going to be modeling a custom-monogrammed coaster!

This should be an easy first project. You'll just be modeling a basic circular plate and cutting a letter into it. This will also be easy to print after you've modeled it.

Open Your CAD Program and Create a New Part

Obviously, the first thing you'll need to do is open up your CAD program. (In this example, like all of the others, I'll be using Solidworks.) When the program is open, you can create a new part. For this example, I'll be using millimeters. If you'd prefer to use inches, just divide those dimensions by 25.4.

Extrude a Circle

To create the main body of the coaster, you should start by extruding a simple circle. You need to select a plane to sketch on for the extrude command. Choose whichever plane makes the most sense to you for that specific part, but keep in mind that you might need to rotate the part when you slice it in order to orient it properly. So under the **Features** tab, find the extrude command in your software and select **Front Plane**.

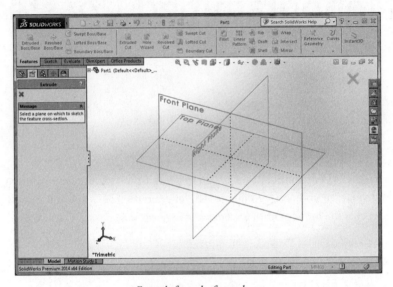

Extrude from the front plane.

The CAD software should then switch you to the sketching toolbar (or it may already be open); in this software, you click the **Sketch** tab. Select the circle icon (defined by the center point) and click on the origin point to place the center of the circle there. You can then just drag the circle out a little bit.

Choose the dimension command in your software (you may need to choose the diameter type specifically) and select the circle. This will place a dimension that defines the diameter of the circle. You can then modify the dimension to fit your needs; in this case, you want it to be 100mm.

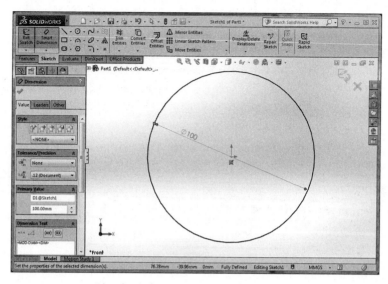

Give the circle a diameter dimension of 100mm.

With the diameter of the circle defined, you can go ahead and extrude it. There should be a button for completing the sketch (in Solidworks, it's a green check mark, as you can see in the following image), which will take you to the feature's options. In those options, give it an extrude distance of 10mm.

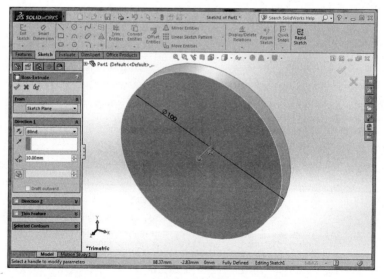

Extrude the circle 10mm out.

Fillet the Top Edge

Next, you want to fillet the top edge of the coaster. This will give it a nice, rounded top. Rounding the top will make it more aesthetically pleasing and will also keep it from having a sharp uncomfortable edge.

Click on the **Features** tab and select **Fillet**. Using the Fillet tool, select the top edge of the coaster. Under **Fillet Parameters** on the left, specify a radius of 5mm. Because the top and side meet at a 90° angle, the edge will be rounded down 5mm on the side and in 5mm on the top. (Remember, fillets are defined by the radius of a circle that is tangent to both surfaces.)

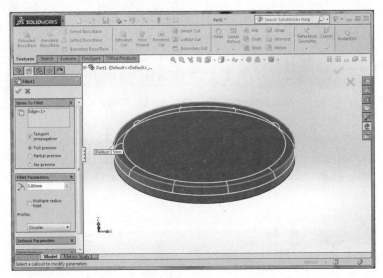

Fillet the top edge of the coaster with a 5mm radius.

Cut the Letter

The final feature to add to the coaster is your initial (or initials) cut out. This will be added with the extruded cut command in your software. After choosing the command in the **Features** tab, select the top surface of the coaster to sketch on.

Before you add the letter, you first need to add a guide to position the text. You could just eyeball it and place it where it looks good, but as I've taught you, it's better to fully define everything. The text (for a single letter) will be 60mm high, so in order to center the letter on the coaster, you need to put a construction line 30mm below the origin.

Start by clicking the line icon to draw a horizontal line below the origin point. You then change that line to a construction line so it's not used for the feature by right-clicking on the line the checking the "for construction" option in your software. Add a dimension to put the line 30mm below the origin point.

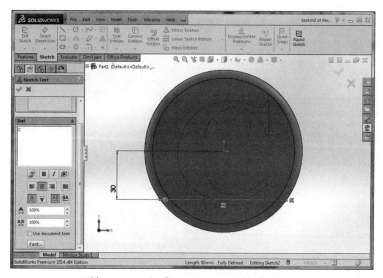

Add a construction line to use as a guide for the text.

You can now go ahead and add the text. In the text command of your software, select your construction line as the guide. Next, change the font options so the text is centered and is 60mm tall. Choose a font style that suits your tastes, and then type in whatever letter you'd like to use.

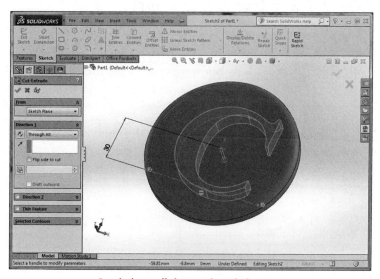

Cut the letter all the way through the coaster.

Finish the sketch to move back to the extruded cut options in the **Features** tab. You can either specify a cut distance of 10mm or simply choose **Through All** (or however it's specified in your software), which will cut through all of the material in that direction. Or you can just cut in a couple of millimeters if you don't want it to go all the way through. Complete the feature, and your coaster will be finished!

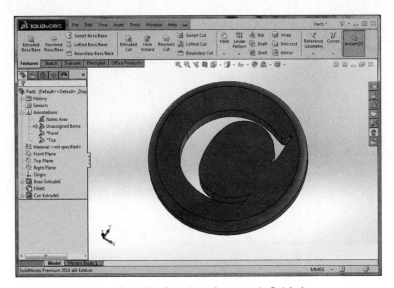

Once the cut has been done, the coaster is finished!

Export the .STL File and Print

If you're satisfied with how the model turned out, you can export it as an .STL for printing. The steps to export the file will depend on your specific CAD software, but generally you'll either "save as" or export as an .STL. When you export it, be sure to choose millimeters, because that's probably what your slicing software will be set to as the default. You also want to save it at the highest-quality settings, because this model has lots of curves in it.

With the .STL file in hand, you can go ahead and print it. This model won't require any special settings to print, so just use basic settings. You can then simply start your print and let it cool afterward, and soon enough, you'll have your very own monogrammed coaster!

Project 5: Monogrammed Coaster   **257**

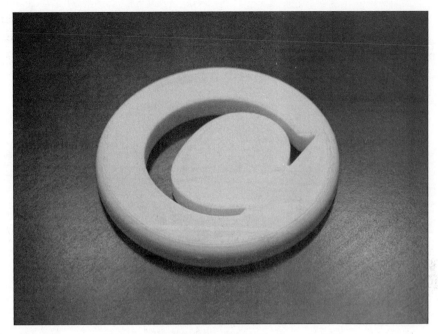

Your printed coaster (if your initial happens to be "C").

PROJECT

6

Custom Storage Drawer

Project Time: 1 hour

Your second modeling project is going to be quite a bit more difficult than the first one. You'll be modeling a custom storage drawer for the storage shelf you printed for project 4. You'll be able to design the compartments however you wish to suit whatever you'd like to store.

Because this will be harder than the first modeling project, it may be beneficial to spend some time modeling some more basic designs on your own to get a feel for the software before doing this. I'll also be assuming that you know the basics of using the commands and software, so I won't be going over each feature in great detail; however, you can check out project 5 if you need a refresher on where to go and what to do in the software. If you're ready, let's jump right in!

Open Your CAD Program and Create a New Part

Start by opening your CAD software and creating a new part. As usual, my recommendation is to work in millimeters. This is especially true for this part, because it will be sliding into the existing storage frame.

The frame was designed in millimeters, so the best way to ensure a proper fit is to use millimeters for the new drawer as well. However, if you prefer to use inches, you can divide all of the measurements by 25.4.

Extrude the Body of the Drawer

For this part, you need to start with an extrusion on the top plane. This is because the main body of the drawer will be created by extruding from the side, and you want it to lay down flat on the build plate when you print it. So in this case, the top plane actually corresponds to the front of the part.

The opening in the frame is a rectangle 25mm tall and 100mm wide, with 5mm fillets on the corners. In order to make the drawer fit into the frame, it needs to be slightly smaller than that opening. You can draw the 25×100mm opening with construction lines, and then offset those inside by .50mm. You can then extrude the offset lines by 100mm (the length of the drawer).

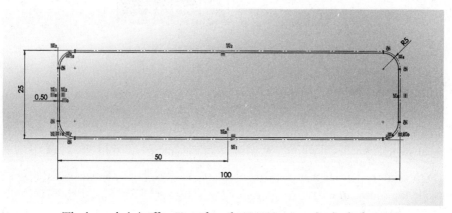

The drawer body is offset .50mm from the 25×100mm opening in the frame.

Cut an Opening for the Handle

The next feature is a cutout that will allow room for the drawer handle. This is so the handle doesn't protrude from the front of the drawer. This design choice is mostly cosmetic, but will also help the drawer to be printable on smaller printers.

To make this feature, you use the extrude cut command in your software on the top of the part (or the front plane). Make an arc with a 100mm radius, with its endpoints 15mm from each side of the drawer. Next, draw a line connecting the endpoints to make it a closed shape. You can then simply do the cut all the way through the part.

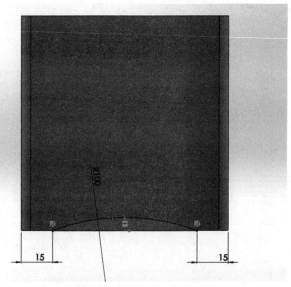

Cut out a place for the drawer handle.

To finish the cutout, add a 50mm fillet to the ends of the cutout. This should be where the endpoints of the arc were. The purpose of the fillets is simply to round the edge and give it a more finished look and feel.

Design the Compartments

This next step is where the customization comes in. You essentially hollow out the drawer and divide it into the desired compartments in one feature. How you divide it up is completely up to you.

However, you should stick to a few basic conventions to make sure the drawer turns out right. In my design, I gave all of the walls a 2mm thickness. This should make it sturdy without being too bulky, so I recommend you do the same.

To do this, start by offsetting the outside lines of the part by 2mm all the way around. Next, draw construction lines to divide it up into the desired compartments. Offset those construction lines by 1mm in both directions (to make the walls of the compartments). Then just trim the overlapping lines so that each compartment is its own fully closed shape.

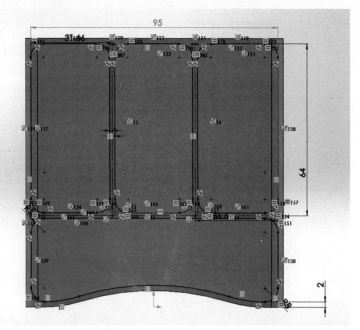

Design the compartments based on your needs. However, it's a good idea to give the walls a 2mm thickness to make them sturdy.

For the options in the extruded cut feature, choose what's used to indicate "offset from surface" in your software. You can then select the bottom face of the drawer. Enter **2.00mm**, and it will cut the compartments down to 2mm from the bottom face.

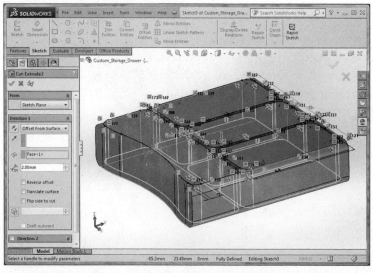

Offset the cut 2mm from the bottom of the drawer.

Add a Rough Handle

Now you can start adding a basic drawer handle. You can use the extrude command in your software, and use the bottom of the drawer as a reference.

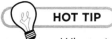

HOT TIP

Why not put the handle in the middle of the drawer? Because doing so would require the use of support material to print the drawer. Extruding it from the bottom will allow the drawer to be printed without any supports, which will improve the quality.

The exact dimensions you use for the handle are up to you. But if you want them to match the other drawers you've already printed, you can use the dimensions shown in the following image. It's a simple rectangle centered on the part that is 30mm wide and protrudes 5mm from the front of the drawer. The outside corners have a 5mm fillet, and the handle itself is 3mm thick.

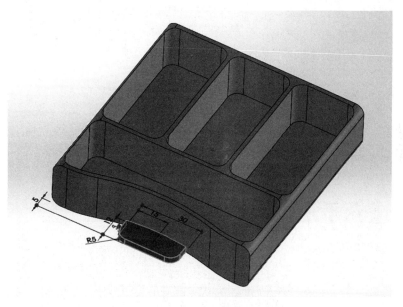

Create a basic handle with the extrude feature.

Add Fillets to Handle

Adding fillets to the handle serves two purposes: to improve the aesthetics and to strengthen the handle. The aesthetic reasons are a pretty obvious: filleted edges just look nicer. But they're also important because they make the handle stronger. The gradual transition takes away the weak point that would otherwise be present at the 90° edges.

In this case, two different fillets are used. On the vertical edges on either side of the handle, 10mm fillets are used. On the top of the handle (the horizontal edge), a 3mm fillet is used. This gives it strength while still leaving a flat spot for you to grip when opening the drawer.

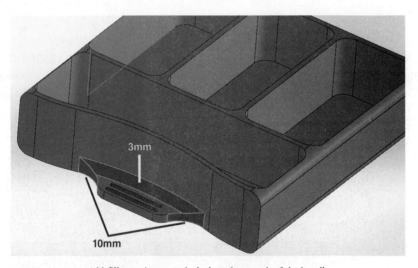

Add fillets to improve the look and strength of the handle.

Add Ridges for Grip

The final two features you'll create are there to make a couple of small ridges on the handle. Their purpose will be simply to make the handle easier to grip when you're opening the drawer.

The first step to making these is a simple extrusion on top of the handle. I made mine by drawing 1mm-diameter circles 20mm apart, and then connecting them with lines on the top and bottom. The insides of the circles can then be trimmed away. The center of the first ridge is 2mm from the edge of the handle, and the second ridge is 3mm from that.

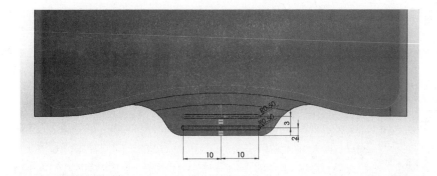

Add a couple of small 1×21mm ridges to the top of the handle.

The final feature is .25mm fillets all the way around the top and bottom edges of both ridges. This is just to smooth out the ridges and make them more comfortable for your fingers.

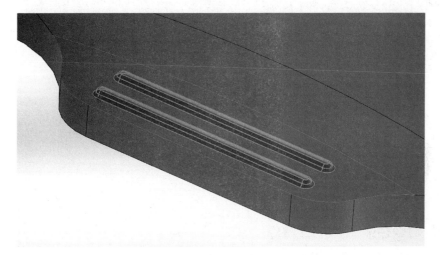

Fillet the top and bottom edges of both ridges.

Export the .STL File and Print

With the part finished, you can export it and print it. Export the .STL file at high-quality settings to make sure you retain the curves of the model. You can then just print it using the same settings you used to print the drawers in project 4 and allow to cool before removing.

Your finished drawer should look something like this, but with whatever compartment layout you decided to model.

PROJECT

7

Dust Collector

Project Time: 45 minutes

A dust collector is a simple little device you feed your filament through. Inside, a small sponge catches any dust that was on the filament. Dust can cause some major headaches in 3D printing, including nozzle clogs, so it's best to avoid the problem altogether by using a dust collector.

This project also gives you the chance to experiment with parts that are designed to fit together, helping you get a feel for the accuracy of your 3D printer and how to design parts that fit snuggly. So boot up that computer and let's get started!

Create a New Part and Revolve the Body

Go ahead and open your CAD software to create a new part in millimeters. The basic design of the dust collector is pretty simple: a hollow cylinder divided into two parts that clip together. This means you'll actually be modeling two different parts.

For the base of the first part, start by using the revolve feature in your software. You can just use the same dimensions that are in the following figure. The only exception is the 1.25mm radius dimension right next to the centerline, which is dependent on the size of your filament. My filament is 1.75mm, so I made the opening 2.50mm to give it some room. If you're using 3mm filament, you'll want to make that dimension something like 2mm radius.

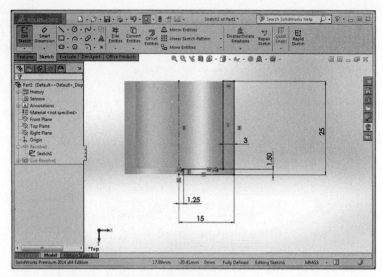

The first feature is a revolve to create the cylindrical shape of the part.

Cut a Groove

Next you'll be cutting a groove on top of the cylinder for the clips to fit into. This is a pretty simple operation with the revolved cut command in your software.

Using the same centerline from the origin you used for the initial feature, you can just draw a circle at the edge of the part. Put the centerpoint of the circle on the edge of the cylinder so it's constrained there, and then just specify a diameter of 2mm and a distance from the top of 3mm.

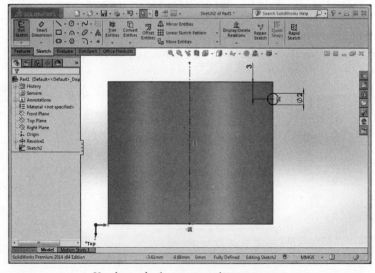

Use the revolved cut command to cut a groove.

You then need to add fillets to the top and bottom of the groove. This will just help the clips fit more cleanly when the two pieces are put together. In the **Features** tab, choose **Fillet** and specify 1mm fillets. Next just choose the top and bottom edges of the groove, and finish the feature. That's all you need to do for the first part, so just save it.

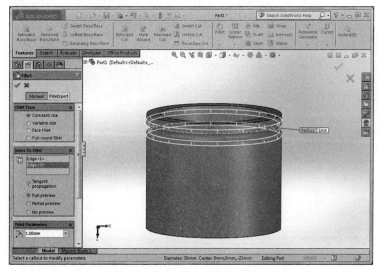

Fillet the edges of the groove you just cut.

Create the Second Part and Revolve the Body

Now it's time to create the second part. This part is a little more complicated, but it starts the same way: with a revolved base.

Just like with the last part, you can use the dimensions from the following image to create the revolve. Again, you may need to change the dimension closest to the centerline if you have 3mm filament.

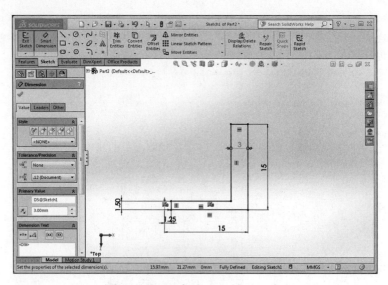

The second part also starts with a revolve.

Make the First Clip

You can't just extrude the clips because you need them to follow the curve of the cylinder in order to function properly. So you should use the revolve command in your software.

Once again, you can use the dimensions from the following image. If you're really paying attention, you may notice it's offset out and up by .25mm compared to where it should be based on the first part. This is to give it a little clearance, so the two parts fit together more easily.

When you actually finish the revolve, don't do it a full 360°. You want the clips to be flexible, so you should only revolve it by a small amount. (I used 30° on mine.)

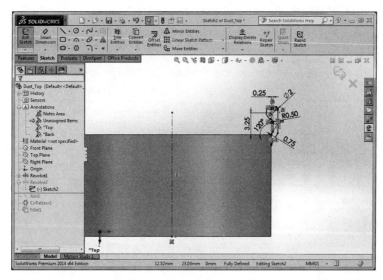

The revolve command should be used for the clip, so it follows the curve of the cylinder.

Copy the Clip

To avoid having to remodel the clip three more times, you can use the circular pattern button in the **Features** tab. But in order to use that command, you need to first create an axis to use as a reference. This is easy enough; just create the axis by selecting any of the cylindrical surfaces.

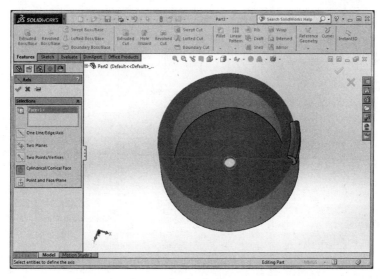

Create an axis to use for the circular pattern command.

With the axis created for reference, you can pattern the clip. Select the first clip, and then select the circular pattern command. Under **Parameters,** select the axis of revolution, type in **90.000** to separate the clips by 90°, and enter **4** for the number of clips (the original plus the three copies).

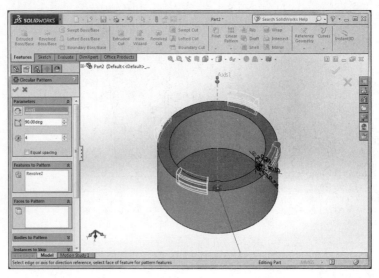

Use your circular pattern command to copy the clip.

Finally, add some fillets to the edges of the clips. This will just make the clips a little bit stronger.

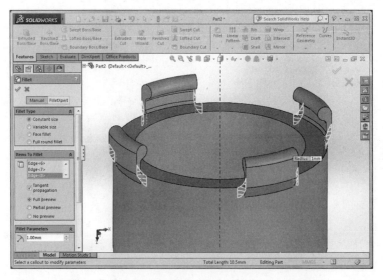

Fillets alleviate stresses on edges and make the part stronger.

 **WATCH OUT!**

You could have added fillets to the first clip and then patterned it along with the rest of the clip, but I've found that patterning multiple features together sometimes causes errors.

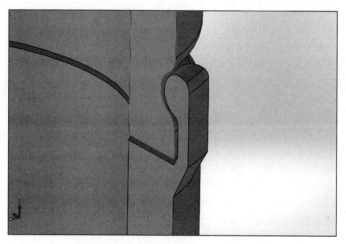

You can see how the parts will clip together in this section view.

Export the .STL Files and Print

As you've done with the previous projects, you just need to export the part as an .STL to print it. The only difference here is that you'll need export both parts. You can then load both parts at the same time and slice them together to be printed. To do this, just load the first part and then the second part. The parts should automatically be placed next to each other for slicing. From then on, you slice the parts the same way as you normally would.

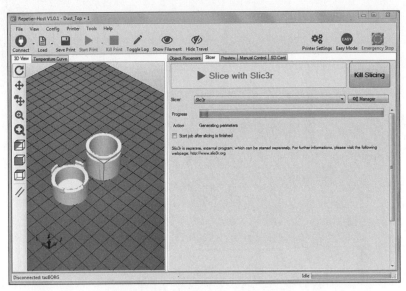

Both parts can be printed at the same time.

Once your parts are done printing and cooled, you can start using them to collect dust. All you have to do is feed the filament through the holes, cut off a small piece of a household sponge and put it around the filament, and then clip the two pieces together. As the filament slides through, the sponge will collect any dust on the part.

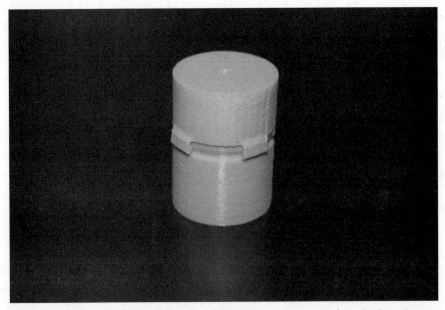

The two parts should be able to clip together. If they don't, you can try tweaking the dimensions to experiment with how parts fit together.

PROJECT

8

Reverse Engineering a Useful Part

Project Time: 30 minutes

This project will involve reverse engineering a part from around your house: an electrical socket cover. Why a socket cover? Because everyone has one in their home, so it's something everyone will be able to follow along with. I'll be modeling a standard U.S. socket cover, but the process should be pretty similar for other kinds of sockets in other countries.

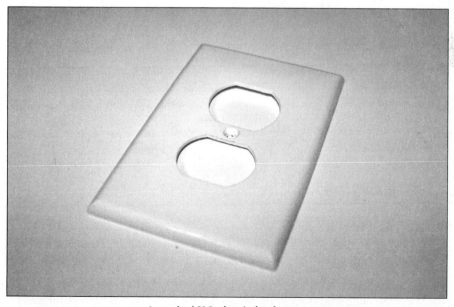

A standard U.S. electrical socket cover.

The idea here is that you'll be learning how to take measurements from the original part. So I won't be providing you with the actual dimensions I used in most cases. Instead, you'll be in charge of coming up with the measurements yourself.

Create a New Part and Extrude the Body

Open up your CAD software and create a new part. As always, I personally prefer to work with millimeters. But you're welcome to use inches if you prefer, especially since you'll be taking the measurements yourself.

You can start by modeling a simple rectangle to get the overall body of the cover done. You want to extrude off of the front plane, so it lies flat on print bed. Just measure the length and width of the cover to draw a rectangle, and make sure it's centered on the origin. Next, measure the overall thickness of the cover for the extrusion thickness.

Model a simple rectangular solid for the body of the cover.

Fillet the Edges

As I'm sure you've noticed, the outside edges of the socket cover are rounded. This is really just a cosmetic feature, but it's one which is easy to add.

You simply need to fillet the top outside edges of the cover. Because it's just cosmetic, you don't need to get the radius of the fillet perfect; just pick a radius that looks right to you.

Project 8: Reverse Engineering a Useful Part

Fillet the top edges to improve the aesthetics. As you can see here, I used a radius of 3mm, but the radius you use is entirely up to you.

Shell the Cover

Unlike the fillets, this feature is important to the functionality of the part. You'll need to use the shell command in your software to essentially hollow out the cover. To do this, you first need to measure the thickness of the plastic that makes up the cover (not the overall thickness).

Measure the thickness of the plastic that makes up the cover.

Once you know the thickness of the plastic, you can then create the shell feature. To do this, you need to specify that thickness you measured, as well as the face you'd like to delete. In this case, you want to delete the back face of the cover (opposite of the fillets).

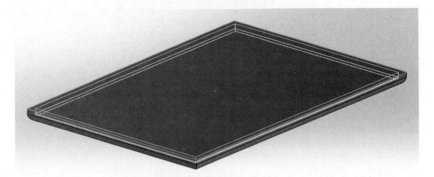

Create the shell by specifying the thickness and which face you'd like to remove.

Cut One Socket Opening

Now it's time to go add an extruded cut for the opening for the actual socket. If you look closely, you'll see that it's actually a relatively simple cutout. It's made up of two straight lines that are connected by two arcs (at least on the U.S. socket).

You can model this in a number of different ways, but I recommend you start with the two straight lines. Draw them based on their length and distance from the top (they should be centered from left to right).

Next, you can draw the arcs. You know their endpoints (the straight lines), so you just need to figure out the midpoints of the arcs. You can do that by measuring the width at the widest point and drawing a construction line centered between the two straight lines to represent it. You can then just draw an arc with its endpoints on the straight lines, and its midpoint on the end of the construction line.

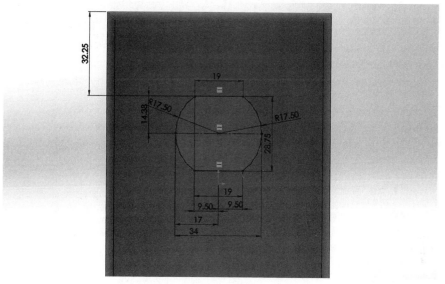

Cut the opening for the electrical socket.

Mirror the Socket Opening

The openings for the electrical sockets are symmetrical, so all you'll need to do to make the second opening is to mirror the first. Select the mirror command in your software and use the top plane as the mirror plane, with the first opening as the feature to mirror.

Mirror the first socket opening to create the second.

Create the Screw Hole Support

On this particular socket cover, there is additional material on the back where the screw hole goes through. This is to add support for the screw, where it pushes against the outlet mounted on the wall. There are also some ribs on the back, but they're probably not integral to the function of the part.

This is a pretty simple feature—just a rectangular support with rounded corners. To begin, measure the length, width, and depth of the support using an actual socket cover.

HOT TIP

Measuring the width and length are easy enough, but the depth is a little harder. If your calipers have a depth gauge at the end, this is one feature where it will come in handy. If you not, you'll just have measure it as best as you can.

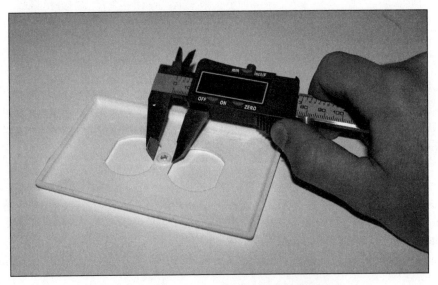

Measure the length, width, and depth of the support.

The corners of the support are rounded. However, the exact radius almost definitely isn't going to affect how well the part functions. So just estimate the radius of the fillets as best as you can.

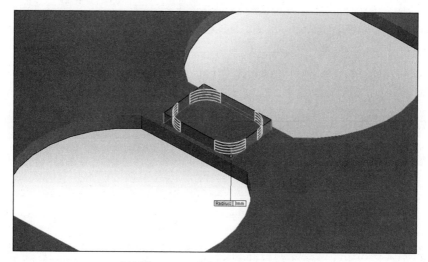

Add fillets to round the corners of the support.

Add the Screw Hole

The screw hole on this part is countersunk to allow the screw to sit flush (or almost flush) with the surface of the cover. To make this, start by measuring the diameter of the actual hole (this is easiest to do from the back) and cut that.

Finally, you need to add the countersink. This can easily be created by using the chamfer command in your software on the edge of the hole. However, you need to know two things: chamfer distance and angle. The distance is easy enough; all you have to do is measure the diameter at the top of the countersink and divide it by 2. The angle will be harder to measure, though. However, 45° is a pretty safe assumption, because it's commonly used for countersunk screw holes.

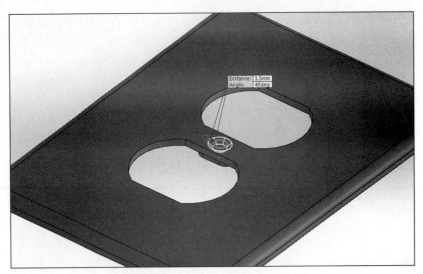

A chamfer can be used to add the countersink.

Export the .STL File and Print

The best way to determine how well you measured and modeled the part is to actually print it out and try it. You can export it the same way as any other part, and I recommend using the highest-quality settings. When you print it, you shouldn't need to use any supports, and you can probably get away with a moderate infill (like 25 to 50 percent). Let the part cool before removing.

The finished cover mounted on the wall, and it fits correctly!

PART

5

Advanced Usage and Techniques

If you've already read the first four parts, you're practically a 3D printing expert already. So in this part, I introduce you to some more advanced techniques and uses for your 3D printing—things like printing with more exotic materials, using multiple extruders, and some popular modifications you can make to your 3D printer. I also tell you a little bit about some uses for your 3D printer that you may not have even realized were possible.

CHAPTER 18

Printing with Other Materials

Throughout this book, I've been talking almost exclusively about printing with PLA and ABS filament. The reason for that is pretty simple: they're by far the most popular filament types on the market. The vast majority of all hobby 3D printing is done with those two materials.

But as popular as PLA and ABS are, there are many other types of filament material on the market. None of these are even close to approaching the popularity of PLA or ABS, but they do have their uses. Most of them are specialty materials designed to be used for particular types of prints. In this chapter, I go over some different materials you can use, as well as what alterations you'll need to make in hardware and settings to print them.

In This Chapter

- Unique filament materials you can print with
- Special hardware needed to print certain materials
- Settings you'll need to change to support other materials

What Materials Are Available?

So what other filament materials are on the market, and what kind of prints are they ideal for? While new filament materials are being developed and released all the time, these are some of the ones which are gaining popularity now.

Nylon

Nylon is a very common plastic type that has been popular for decades. It's recently become available in filament form for 3D printers. The desirable properties of nylon are its strength and its low coefficient of friction.

Parts that are part of a moving system, such as bushings and slides, are usually made from nylon because of its low friction. So nylon filament is generally used for the same purpose: to print parts where friction is a concern.

Polycarbonate

Another material that has been around for a very long time is polycarbonate. Polycarbonate is a very strong material, making it ideal for applications where durability and impact resistance are important. For example, bulletproof glass is made from laminated layers of polycarbonate.

WATCH OUT!

Please do not attempt to shoot something you've printed in polycarbonate. It's very strong, but bulletproof glass is made using a very specific manufacturing process, and regular polycarbonate will not stop a bullet. So go ahead and forget any ideas about 3D printing bulletproof armor or something.

So if incredibly high strength is required in a 3D printing application, polycarbonate is a good choice. In fact, it's probably the strongest filament currently available for consumer 3D printers.

Flexible Filament

I've talked about flexible filaments a few times already, so you're probably already at least a little bit familiar with the idea. Flexible filament is a generic term used to describe any of the flexible, rubbery filament materials available from a handful of manufacturers. The actual formulas used vary from manufacturer to manufacturer, so these are really grouped by their properties (not because they're a particular kind of plastic).

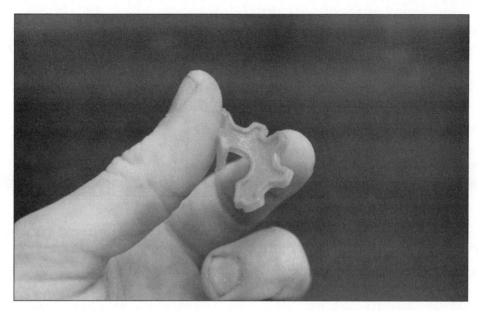

This part was made from NinjaFlex, a popular brand of flexible filament.

Flexible filament has grown in popularity tremendously over the past couple of years. In the hobby market, it's often used for things like robot or RC tires, grips, and anything else that requires a rubberlike feel. With a dual extruder setup, it's popular to print both flexible filament and a traditional hard filament in the same part. For example, you could print a wheel and tire as a single piece from the two different materials.

Wood Filament

One filament material that is pretty surprising and unusual is "wood" filament. There are only a couple of manufacturers currently making this filament, so it's quite unique. The material itself is essentially a mix of polymer and wood fiber.

When printed, the material actually looks remarkably like wood. In fact, it can even be sanded just like wood. The appeal is obvious: you can 3D print objects that look like they're made of wood.

PET

Polyethylene terephthalate (PET) is another plastic that has been around in various forms for decades. It's one of many plastic types that are commonly used in a wide range of products, from bottles to tapes. As far as mechanical properties go, it's fairly similar to other filament materials on the market.

The primary advantage that PET has over other filaments is how well it prints. It has strength similar to ABS but doesn't require a heated bed and doesn't warp much (if at all). Basically, it combines the advantages of PLA and ABS while avoiding their disadvantages. However, PET has only recently been made available in the form of filament. It's only being made by a couple of manufacturers, so it's still a lot more expensive than PLA and ABS.

HIPS

High-impact polystyrene (HIPS) has traditionally been used for plastic parts that don't require a lot of strength (like packaging), but it has gained a new use in 3D printing as a support material. In a dual-extruder 3D printer, HIPS can be used as the support material while another material is used for the actual part.

The reason that HIPS is well suited for this application is because it dissolves in limonene. The primary material doesn't dissolve in limonene, so this is ideal because the support material can be removed chemically instead of with a pair of pliers. While there are other filament materials that are also suitable for this task, right now HIPS is the most common.

Hardware Needed

It's probably not surprising that some of these materials will require special hardware. As new materials are added to the market, it's likely that the hardware requirements will become even more specialized. I can't cover the required hardware for all of the filament materials, so instead I'll give you an idea of what kind of hardware is commonly needed.

All-Metal Hot Ends

All-metal hot ends, as I've touched on previously, are simply hot ends which are constructed entirely from metal. Traditional hot ends are made from a combination of metal and plastic, which is adequate for standard hot end temperatures. But a lot of these materials require that the hot end be much hotter than is necessary for ABS or PLA.

Some of these materials needed to be printed at temperatures of 300°C or higher. That's simply too hot for traditional hot end constructions, which can become damaged at temperatures that high. All-metal hot ends can handle these extreme temperatures without becoming damaged.

HOT TIP

All-metal hot ends are becoming very popular because of their high-temperature capabilities. But they're not perfect for all situations. Many all-metal hot ends are prone to clogging and jamming, so it's usually not worth adding one to your 3D printer unless you know you need to print at high temperatures.

Print Fans

I've already talked about how important print fans are in Chapter 10. They're very helpful for some features, like bridges and small layers. But some materials require a print fan in order to print at all.

Print fans exist to quickly cool the filament after it has been deposited onto the bed or the previous layer. Materials like ABS solidify pretty quickly on their own, so a fan isn't necessary. But other materials are more finicky and have to be actively cooled in order to solidify quickly enough.

Of course, print fans in general are a good idea anyway, so it's debatable whether they should even be considered "specialty hardware." But strictly speaking, they aren't necessary for all prints and materials, making them optional. It's also worth mentioning that the results from a fan alone aren't always good. Sometimes it's necessary to add a fan shroud to direct the cool air in order to get good results.

The Lulzbot TAZ comes with a shroud for the fan to direct cool air just below the nozzle.

Heated Beds and Bed Materials

You already know that heated beds are required for printing ABS, but there are other materials that require a heated bed as well. Plus some materials like nylon even benefit from special build platform materials. Traditional adhesion improvement methods don't always work for all materials. Therefore, some materials have their own unique techniques for making sure the filament sticks to the bed.

 FASCINATING FACT

Phenolic resin plastic was the very first commercially available synthetic resin. It was originally sold by the Bakelite Corporation, and was immediately popular in a wide range of products. Since its invention, it has been used to make virtually everything plastic at one time or another.

At first, that may sound trivial, but it can actually be a fairly big deal for some materials. For example, nylon filament reportedly adheres best to phenolic resin (usually known by the trade names Garolite and Bakelite). Different filament materials have different ideal bed materials, and it's often helpful to switch between bed materials depending on the filament you're using. For instance, a glass bed works well for ABS and PLA, but you may want to switch to Garolite for nylon.

Printing Techniques

As you probably surmised already, all of these filament materials don't just require special hardware. In addition to that hardware, the different types of filament also require specific settings. Just like how you have to adjust the settings when you switch between PLA and ABS, you have to adjust the settings to use these more exotic filament materials as well.

Temperature

The print temperature is the most obvious setting that will pretty much always need to be adjusted. Almost all filament materials have their own ideal temperature setting, and the range can be 100°C or more. The filament manufacturer should have some information on what temperature is ideal for their filament, so be sure to check with them first. For example, PLA is usually printed around 200°C, ABS at 230°C, nylon at 250°C, and so on. If you tried to print at PLA temperatures, it wouldn't even extrude. PLA printed at nylon temperatures would burn or pool on the bed.

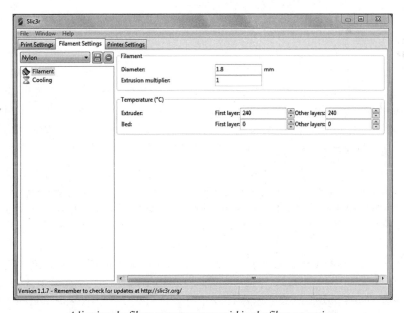

Adjusting the filament temperature within the filament settings.

Once you have an idea of what the temperature should be, it's best to experiment a little to find what gives you the best results. That particular filament may print better on your printer when it's a little cooler or hotter than recommended. Print a few small parts to get your temperature dialed in, and then save it as a configuration for that material.

You should also follow the same steps for the temperature of the heated bed. Some materials won't need the heated bed to be on at all, while others might need it to be very hot. However, the temperature of the bed is much more forgiving than the temperature of the hot end, so as long as you're within 10°C or so, it should work just fine.

Speed

Print speed is another major variable when you're switching between filament materials. Many of the more exotic materials have to be printed very slowly and will produce poor results if they're printed too quickly. This is basically determined by how quickly that particular material can melt in the hot end, which can be dramatically different between materials. For example, PLA can usually be printed at very fast speeds because it flows well. Nylon, on the other hand, usually needs to be printed slowly (around 40mm/s) or it won't bond properly to itself.

Just like with the temperature settings, you should start by looking at what the manufacturer recommends. If that information isn't available, do some searching to find out what settings other people are using for the same type of material. Once you've got some numbers to start with, you should fine-tune it by printing a few parts, just like you did after adjusting the temperature settings.

Cooling

As you learned earlier in this chapter, some materials will require active cooling while others shouldn't be cooled at all. Once again, this information can usually be found from the manufacturer of the filament or by searching around to see what others are doing. But cooling can be a lot more subjective than temperature and speed.

For some materials, you'll want the fan on at full blast all the time. For others, it shouldn't be turned at all. There is also a lot of middle ground when it comes to using a fan. You can have the fan come on just for certain features (like bridges or overhangs), or the fan could be on a low speed. For example, PLA virtually requires a print fan to turn out well. However, ABS should only have the fan on for extreme bridge or overhangs; otherwise, it will warp. As a general rule of thumb, to start your experimenting, the fan should be off for materials that are prone to warping, while materials that don't have this issue can usually be printed safely with the fan on. Getting your cooling settings right will definitely take some patience and experimentation.

So how do you get it adjusted properly? Start by running a print with the fan on high the entire time. You can then try running another print with it off completely. Compare the results and see which one worked better. Once you know how the material reacts to cooling, you can start adjusting the settings for particular features. For example, you might have the fan off completely for the first few layers, on low for the rest of the print, and on high for bridges and overhangs. The point is, you'll need to look at the results to see what works best.

Layer Thickness

This is a setting you may not expect to have to change, and in most cases you won't. But a small minority of filament materials (such as nylon) can be a bit sensitive when it comes to layer height/thickness. Unless the filament manufacturer recommends otherwise, it's best to start by using whatever layer height you would normally use.

But if you're getting poor results (especially inconsistent extrusion and poor surface quality), layer thickness may be the culprit. When this happens, it's almost always because the layer height is too small, so try increasing it a little bit and see if that improves your results. In extreme cases, you may even have to switch your nozzle out for some materials that simply don't flow well through a small nozzle.

The Least You Need to Know

- The options for filament materials are rapidly increasing, and there is now a pretty large variety of materials available on the market.
- Some materials are better suited to particular prints than others. You should pick the material based on your needs for specific parts.
- Many materials require specific hardware—such as all-metal hot ends, print fans, and heated beds and bed materials—in order to print. Before you order a new filament type, be sure to check that your 3D printer is capable of printing it.
- The print settings will almost always need to be adjusted for new materials. Be sure to take the time to tweak your settings until you're getting good results with a new material.

CHAPTER 19

Modifying Your Printer

Once you're comfortable with your 3D printer and have gotten bored with printing pencil holders and coasters, you'll probably start thinking about how you can upgrade or modify your 3D printer. There are an endless number of modifications you can do to your printer to make it bigger and better, and if you're the tinkering type (which you probably are if you're reading this book), I'm sure you're eager to get your hands dirty.

You can really modify everything on most printers with enough work. After all, you have a 3D printer you can use to make new parts for upgrades. But there are a handful of popular modifications that you can do to most 3D printers to improve their functionality. So in this chapter, let's take a look at some of the more common upgrades and modifications you can do!

In This Chapter

- Adding fan shrouds or heated beds
- Making the switch to all-metal hot ends
- Installing extruders and extending axes
- Modifications for milling or laser cutting
- Sites for purchasing parts or finding 3D models to upgrade your printer

Adding a Fan Shroud

If your 3D printer doesn't already come with a fan shroud, adding one is definitely the most popular 3D printer modification. In fact, it's usually the first modification people make to their 3D printers.

WATCH OUT!

Attempting to modify your printer always carries with it the risk of damaging the printer. It's also likely that your warranty (if you have one) will be voided by the modifications. Therefore, make any modifications and upgrades with the risks in mind.

While many manufacturers include a print fan with the printer, it's often just a small fan blowing in the general direction of the nozzle. Fan shrouds improve the situation by acting as a funnel to direct and focus the air just below the nozzle. Doing so usually improves the cooling capabilities of the fan quite dramatically. While you shouldn't expect miracles, a fan shroud can definitely make a difference.

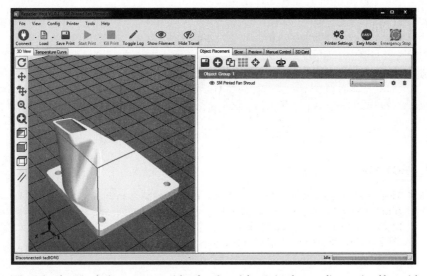

The Printrbot Simple does not come with a fan shroud, but Printrbot supplies a printable model to add one yourself.

Adding a fan shroud is usually a fairly easy job. Most 3D printer models on the market have fan shrouds already modeled and available for download. In many cases, you'll be able to attach the shroud using the mounting screws that are already on the printer for the fan.

So all you need to do is find a fan shroud model for your 3D printer model and then just download and print it. Once it's finished printing, you can just mount the shroud to the fan. You should immediately see a cooling improvement.

Adding a Heated Bed

If your 3D printer didn't come with a heated bed, your material options are going to be pretty limited. PLA will be the only commonly available material you can print successfully. Now PLA is actually a really great material that prints well and is reasonably strong, but it's always nice to have the option to use other materials.

To do that, you need to add a heated bed to your 3D printer. The difficulty of doing this depends on three major factors: how large the bed is, the control board you're using, and the power supply you're using.

Bed Size

Generic 3D-printed heated beds usually come in a relatively small selection of standard sizes. If your bed size is something common like 6×6 inches, finding a heated bed shouldn't be difficult. But if you're looking for something strange like 3×9 inches, you're probably not going to have much luck finding one.

It is possible to make your own heated bed in whatever size you like, but it's not easy to do. You'd essentially have to make your own printed circuit board (PCB) at the size you need, and you'd have to make sure the traces were the correct size to get the resistance right. Or you can contact a heated bed manufacturer and have them make you a custom bed, but that will be a little pricey.

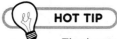

> **HOT TIP**
>
> The heated bed doesn't have to be exactly the same size as the build platform. As long as it covers the print area (or is close to it), it should work fine. For example, if your print bed is 150×150mm, a 6×6-inch heated bed (152.4×152.4mm) would probably work just fine.

However, that shouldn't be an issue for most people. 3D printers generally come with pretty standard bed sizes, so finding the right size for your printer shouldn't be too difficult.

Control Board

In order to add a heated bed, your control board has to have an output to power it and an input for the thermistor (which measures the bed's temperature). Most modern control boards will have the output and input for the heated bed, but not all will.

If your control board doesn't have the connections for the heated bed, you'll need to upgrade to a control board that does. There are many control boards on the market that should be compatible with just about any 3D printer. Changing control boards will take a fair amount of rewiring work, but the process is usually well documented on websites like RepRap.org.

If the control board already has the heated bed connections, adding the heated bed will be significantly easier. However, if the bed is large (over 8×8 inches), you'll also need to make sure it can handle the current required for the bed. For example, if the bed draws 15 amps and the control board can only handle 10 amps, you'll have a problem. But that should only be an issue for large heated beds.

Power Supply

The final concern when adding a heated bed is whether or not the power supply can handle it. Heated beds use a lot of power, so the power supply has to be able to provide the power for the heated bed as well as the rest of the 3D printer. It's not uncommon for the stock power supply to be inadequate once a heated bed has been added.

Luckily, upgrading the power supply is a pretty easy job. Power supplies come in a variety of styles and form factors, with all kinds of different connections. Some printers use laptop-style power supplies, which means you'll need to make sure the jack is the right size and shape (in addition to it meeting the power requirements).

 **WATCH OUT!**

Rewiring your printer, especially when it comes to the power supply, can be potentially dangerous. It's possible to improperly connect the wires, which could result in electrocution or fire. Be sure you know what you're doing and use the proper safety precautions when doing this kind of work.

Other printers use generic power supplies that have bare connectors you simple screw wires into. Those are the easiest to replace, because you just have to get a more powerful one and switch the wires over. They also tend to be relatively inexpensive.

Many people also use desktop computer ATX power supplies. They're ideal because they're cheap, readily available, and can put out a lot of power. However, they usually have to be jury-rigged to work with 3D printers because they have built-in switches and safety mechanisms. But they're a great option for supplying a lot of power on a budget.

If the new power supply doesn't have a power connector that matches the original, you'll need to do some rewiring for this modification as well. Luckily, this is usually a simple matter of cutting off the old connectors and installing new matching connectors. Most power supplies will only have two wires (positive and negative) that are color coded (black and white or red), which just connect to the corresponding colors on the other end.

A standard ATX computer power supply converted for use with a 3D printer.

Switching to All-Metal Hot Ends

In the previous chapter, I talked about how all-metal hot ends are required for some materials because of their high temperature requirements. For that reason, it has become very popular to switch out the standard hot end for an all-metal model. Making that switch is usually pretty easy to do, but like all things it can become complicated.

Manufacturers of aftermarket all-metal hot ends purposefully make them so they're easy to add to your printer. Obviously, it's in their best interest to make them as universal as possible. But they're still not quite plug-and-play.

There are two factors here: physically making the connections and mounting the hot end. Making the connections is usually as simple as just splicing four wires (two for the heating element and two for the thermistor). The smaller wires will be for the thermistor, and the polarity won't matter. However, you'll usually need to reconfigure your firmware for the new thermistor as well (which entails changing the thermistor model).

Actually mounting the hot end can be a little more complicated. There are a few somewhat standardized mount types, and then a handful of completely unique ways of mounting hot ends. If your printer's hot end is one of the standard mounts, you can often just swap out the hot end. In some cases, you may need to first print some kind of adapter if the mount types are standard but different.

It only becomes a serious headache if your printer has an unusual type of hot end mount. Some manufacturers use completely proprietary hot ends that have unique mounts. In a case like that, you'll either need to find an adapter (or entirely new extruder carriage) that's already been modeled by someone else or model your own.

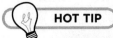

HOT TIP

> Hot end compatibility is definitely something you should look into before ordering an all-metal hot end. Do some searching on the forums for your 3D printer and see what other people are doing. Is there a specific model of hot end that people prefer? Maybe some fit better than others and are easier to mount. If so, that can help you decide what hot end to order.

Installing Multiple Extruders

Adding an additional extruder to your 3D printer (or a few additional extruders) is a very exciting modification. It opens up a whole new set of possibilities when it comes to what you can print. You can print in multiple colors, in different materials, or with a dedicated support material. But it's also a pretty difficult modification to make.

Some 3D printer manufacturers have kits available to upgrade your printer to a dual extruder setup. For example, Lulzbot has an upgrade available for dual extruders. But I'm not aware of any that currently have kits available for adding more than one additional extruder. And it's not particularly common for them to even sell those.

This is a drop-in dual extruder developed by Lulzbot for their TAZ 4 3D printer.

For most people, adding one or more extruders ends up being a completely custom job. You'll need to design a new extruder carriage with mounts for the extruders, upgrade your control board to one with multiple extruder connections, and reconfigure your firmware. None of that is easy to do, and can be quite complicated. The RepRap forums (forums.reprap.org) are a good place to look for information on doing this kind of extensive modification.

Fortunately, the hobby 3D printing community is very good about sharing information and designs. If your 3D printer model is fairly popular, there is a pretty good chance that someone has already done all of that work for you. With a little searching and some luck, it's entirely possible that you'll be able to find the models you need to print, along with information on the control board and firmware.

But even if that information is available, you need to be very comfortable with modifying your printer. You need to take things apart and put them back together, rewire the control board and extruders, and so on. Adding extruders definitely isn't a job for the faint of heart.

Extending Axes

While I'm on the topic of difficult modifications, let's go ahead and talk about extending your printer's axes. It should be readily apparent why this is a desirable modification: it makes your print area larger. Who wouldn't want the option of printing bigger objects?

On the surface, this seems like a relatively simple job: just get some longer smooth rods and a larger build platform. And in some cases, it is that simple. However, it often turns into much more than that.

Those longer smooth rods may force you to modify the frame to fit them. If they're especially long, they may droop, which means you'll need to get thicker rods. Thicker rods will require new mounts and bearings. The additional weight may mean you'll have to upgrade the stepper motors and possibly even the control board and power supply.

As you can see, this often results in a cascading effect. Some printers (like Printrbots) can have their axes extended easily and inexpensively because of the way they're designed. There are even kits available for some Printrbot models to do just that. Other printers (especially those with fully boxed frames) will require a lot more work.

As usual, you should always start by looking at what other people have done. If others have been able to successfully upgrade their printers with extended axes, look at how they did it. If no one else has done it, that's a bad sign. In some cases, it may not even be practical or even possible to do. In cases like that, it's better to just purchase a larger printer.

Converting to a PCB Mill

Earlier in this book, I spent a fair amount of talking about what CNC mills do. If you recall, CNC mills make parts by cutting into a block of material. A printed circuit board (PCB) mill is basically just a type of CNC mill that's designed to cut traces out of standard PCB sheets for electronics projects.

HOT TIP

Many people convert their CNC mills into 3D printers, but going the other way doesn't usually work as well. The components used to build 3D printers simply aren't as robust as those used for CNC mills. This means a 3D printer converted into a CNC mill usually won't have the power or rigidity to mill anything but very soft materials (like plastic, machinable wax, and wood).

Being able to make PCBs at home is a pretty big deal for electronics hobbyists, making the conversion from 3D printer to PCB mill ideal. So if you're interested in making PCBs, converting your 3D printer can be a worthwhile solution.

Converting a 3D printer into a PCB mill is very similar to converting it into a CNC mill since PCB mills and CNC mills are very similar. But the conversion to a mill strictly for PCB use is usually a little bit easier. PCB mills just need to be able to cut through the thin sheet of copper on a blank PCB, which doesn't require a lot of force.

The low power requirements of PCB mills mean they're easier to convert a 3D printer into. The spindle doesn't have to be high power (a normal rotary tool will usually work), and the frame doesn't need to be incredibly rigid. Overall, it's a much cheaper and easier conversion.

The easiest way to accomplish this is to design a Dremel mount for your 3D printer. Ideally, you'll replace the extruder mount with the Dremel mount. That's all the physical modification that is really needed for the conversion. A Dremel (or other rotary tool) has all of the power needed for PCB milling.

Alterations for Laser Cutting

Maybe you're thinking to yourself "Milling is so twentieth century; I want something more futuristic!" Well, if that's what you're thinking, I have a treat for you: lasers! That's right, you can even add lasers to your 3D printer.

Laser cutting is actually a fairly mature manufacturing process at this point and has been around for quite some time. The basic process is simple: a flat sheet of material is laid out and a high-powered laser burns through it to create flat parts. Many parts are created this way, and it's especially popular for flat parts made from wood, acrylic, and various metals.

A laser cutter only needs to move in two axes, so a 3D printer is easily capable of the necessary movements. The conversion is mostly a matter of slapping a laser on where the extruder would go and replacing the bed with a platform designed for laser cutting. The platform can be as simple as a tray with water in it, just to keep the laser from damaging whatever is below the material being cut.

The physical modifications would be very similar to those needed for PCB milling. Just remove the extruder mount and replace it with a mount which holds the cutting laser. The platform below needs to be capable of reflecting the laser, so that it doesn't cut through.

But before you start daydreaming about opening a metal fabrication shop, you should be aware of the requirements for the laser. Cutting through metal requires a very, very powerful laser. Those lasers are expensive, and so is the equipment to adequately power it.

However, some other materials can be cut with relatively inexpensive lasers. Acrylic, for example, doesn't require a particular high-powered laser. This is mostly due to how well it absorbs the energy from the laser (as opposed to metals, which reflect a lot of the energy). Therefore, it's possible to convert your 3D printer into a laser cutter capable of cutting ¼-inch-thick acrylic for just a few hundred dollars.

> **HOT TIP**
>
> If you're interested in converting your 3D printer to a laser cutter, be sure to research what kind of laser is needed for the materials you'd like to cut. Different materials (and thicknesses) all require lasers of varying power. Some materials can be cut with relatively low-powered lasers, while others require very high-powered lasers.

Finding Parts

For most of these modifications, you'll need to be able to source the necessary parts. Finding the specific parts you need isn't always easy, but with the power of the internet, you shouldn't have too much trouble. These days, there are many websites selling 3D printer parts, and a simple Google search will help you find a plethora of them. However, the following are some links to get you started:

McMaster-Carr (mcmaster.com): This is a great resource for all kinds of parts and components. They have an enormous selection of materials, fasteners, and all kinds of machinery components. They're very popular in the engineering world, because their selection is just so comprehensive.

eBay (ebay.com): You can find all sorts of 3D printer parts through this auction and shopping site. This is especially true for inexpensive parts coming directly from China. You would be hard pressed to find better prices anywhere else, though you'll have to be patient if your parts are being shipped from overseas.

Thingiverse (thingiverse.com): This is currently the largest repository of 3D-printable models. Aside from all of the normal knickknacks and decorative items to print, you'll also find a lot of models for upgrading 3D printers. There is a very good chance you'll find a handful of upgrades for your printer that people have already modeled and tested here.

The Least You Need to Know

- There are many modifications you can make to upgrade your 3D printer, some of which are more difficult and expensive than others.
- Adding a heated bed allows you to use materials beyond PLA, making it a popular modification.
- With multiple extruders, you can print in multiple colors, in different materials, or with a dedicated support material.
- McMaster-Carr and eBay are both good resources for finding 3D printer parts online. Thingiverse is the largest online repository of printable 3D models. You can often find models there for upgrading your printer.

APPENDIX A

Glossary

ABS Acronym for acrylonitrile butadiene styrene, a type of thermoplastic commonly used for 3D printer filament. It is strong but prone to warping and cracking during printing.

ABS juice Also called *ABS glue*, a solution of ABS plastic dissolved in acetone. It makes a very sticky substance that can be used as a surface treatment on the print bed to improve adhesion.

acetone A common household and industrial solvent. Acetone breaks down styrene, which allows it to dissolve ABS. It can be used for smoothing ABS prints and for making ABS juice.

additive manufacturing The process used by all 3D printers, in which a part is created by slowly adding layer after layer of material.

Arduino An open-source platform for developing and prototyping electronics. Arduino models are generally small circuit boards without inputs and outputs for controlling various electronics. The Arduino Mega is used by the RAMPS control board for 3D printing.

Basic Input/Output System (BIOS) The firmware interface used on most computers. It's the first thing to load as soon as your computer starts up and controls how the operating system (Windows, for example) is booted.

borosilicate Also known as *Pyrex* (one of its trade names), a type of glass that is formulated to reduce thermal expansion. Because it's much less prone to thermal expansion than normal glass, it's at a much lower risk of thermal shock. Thermal shock can crack traditional glass if it's quickly heated or cooled unevenly, a problem that borosilicate doesn't experience.

Bowden A general category of cold end where the filament is fed from a stationary location through a feed tube and into the hot end.

calipers   A common measurement tool used in a wide range of fields and industries. They come in both analog and digital varieties and are used to measure lengths, distances, and depths. Calipers are capable of very high precision and commonly come in 6- to 12-inch sizes (though larger and smaller ones exist).

Cartesian   An adjective used to describe things related to René Descartes, who was a French mathematician and philosopher. His many contributions to mathematics were the reason for the Cartesian coordinate system being named for him (though he wasn't solely responsible for its development). In the context of 3D printing, it is a way of defining points in 3D space by their X, Y, and Z coordinates.

chamfer   A simple beveled edge that connects two surfaces. If the surfaces meet at 90 degrees, a standard chamfer will cut across at 45 degrees for symmetry. However, a chamfer does not have to be symmetrical and can cut across at other angles as well.

cold end   The part of the extruder assembly that has a stepper motor and drive system to feed the filament into the hot end.

computer-aided design (CAD)   The process of designing parts using computer software (either in 2D or 3D). It also refers to the software used to accomplish that.

computer-aided manufacturing (CAM)   A term used to describe manufacturing processes that are controlled by a computer (generally CNC mills). It also refers to the software used by the computers for controlling the machinery.

computer numerical control (CNC)   A way of controlling machine tools using a computer, often used for computer-controlled milling machines.

control board   The brain of the 3D printer, the control board is responsible for handling and processing all of the inputs and outputs on the printer.

coupler   A mechanical fastener used to connect the lead screws on the drive system to the shaft of the stepper motor.

Delrin   A trade name for a type of low-friction plastic.

digital light processing (DLP)   One of many 3D printing processes, DLP uses projected light to cure photopolymer resin in layers.

direct drive   A type of extruder that doesn't use a gear reduction system. Because of the relatively low torque, direct drive cold ends are generally only suitable for 1.75mm filament.

direct feed   A category of cold end where the drive system is located directly above the hot end, so filament is fed directly from the cold end into the hot end.

drafting   The process of creating technical drawings of parts and assemblies. A person who does this is a drafter (or draftsman or draughtsman). Traditionally, this was done with a pencil and paper, but it is now done almost exclusively with CAD software.

enclosure   A case that surrounds the 3D printer to keep hot air in and cold drafts out. Enclosures can significantly reduce warping and cracking on printed parts.

end stop   A switch at the end of each of the 3D printer's axes that tells the control board when each axis is at its limit.

extrusion   The process used by FFF 3D printers to melt plastic and deposit it by squeezing the molten plastic through a nozzle.

filament   The material used by FFF 3D printers, which comes in the form of a long strand of plastic and is generally wound on a spool.

firmware   The program that runs on the control board for controlling the 3D printer.

flatness   One of many terms used in engineering to specify geometric tolerances. In the case of 3D printing, it describes how flat a surface is.

flexible filament   An umbrella term for any filament material that is flexible, rubbery, and squishy. This can be used as an alternative to the hard and rigid plastic that is more commonly used for 3D printing. The exact composition differs depending on the manufacturer, but the resulting material is fairly similar with all of them.

fused filament fabrication (FFF)   Also called *fused deposition modeling (FDM)*, the 3D printing process used by almost all consumer 3D printers.

G-code   A programming language that 3D printers and other computer-controlled machine tools can use for instructions. The G-code is what is used to give most 3D printers the commands they follow to produce parts.

green sand   A specially formulated material used for sand casting. Sand casting is used to produce metal parts by pouring molten metal into a mold made of sand. The mold is normally produced by forming it around a positive master part; however, 3D printing the mold removes the need to first create a master part.

Hall effect   Discovered by Edwin Hall in 1879, refers to the tendency for voltage in a circuit to change when exposed to magnetism. It can be harnessed to create proximity sensors and is sometimes used for end stops and probes.

heated bed   A type of 3D printing build platform that is electrically heated in order to aid adhesion and reduce warping.

heated build chamber   Similar to an enclosure, except it is actively heated. Heated build chambers can often eliminate warping and cracking entirely.

heating element   A simple electrical resistance-based heating device usually used to heat up the hot end.

HIPS Acronym for high-impact polystyrene, a type of plastic that can be used as support material for 3D printing. It dissolves in limonene, so it can be removed chemically to avoid damage to the printed part.

host software The software that runs on your computer and connects your computer to the 3D printer. It can be used to manually control the printer, as well as to send files to be printed.

hot end The part of the extruder assembly that heats and melts the filament that is fed by the extruder assembly. As filament is pushed into the hot end, a heating element heats the hot end. The temperature is hot enough to almost instantly melt the plastic into a very viscous fluid, which is then squeezed out of the nozzle and deposited on the print bed.

hypotenuse The longest side of a right-angled triangle. It is always the side whose endpoint doesn't touch the right angle.

intellectual property Any idea that is legally protected. Providing legal protection for intangible things like ideas is a complicated matter, and the laws vary from country to country. But in the United States, intellectual property like inventions, music, copywriting, patents, and so on are legally protected property.

layer height How thick each individual layer is. The height is inversely proportional to both quality and print time, so the thinner each layer is, the longer it will take to print and the better the quality will be.

LCD controller A device that connects to the control board and lets you control the 3D printer without having a computer attached.

lead screw A precision-machined component for translating rotary motion into linear motion.

mechanical advantage The amplification of force with the use of a tool or mechanical system. This is usually achieved by trading movement distance for force. A lever is the most basic example of this, because if one side of the fulcrum (pivot point) is twice as long as the other, it will double the force exerted (though it will also double the distance it needs to be pushed). This same basic concept is applied in a vast array of machines using things like gears, pulleys, screws, and so on.

MultiJet Printing (MJP) A 3D printing process that uses multiple nozzles to spray binder onto powder.

nozzle The small opening at the end of the hot end that determines how thick the extruded filament is.

nylon A type of plastic that is noteworthy for its very low coefficient of friction. It can be found in filament form for 3D printing.

open source A philosophy and usage rights system that allows information to be shared freely.

parametric A 3D modeling system used by CAD software that defines the model by a series of parameters which can be modified.

PET Acronym for polyethylene terephthalate, a type of plastic that can be used as a 3D printing surface in film form or as a 3D printing material in filament form.

photopolymer resin A type of liquid resin that solidifies into plastic when exposed to light (usually in the ultraviolet spectrum). Manufacturers can produce the resin in many varieties, with different mechanical and chemical properties.

pitch The distance from one thread to the next. This is what determines how far the screw and nut will move (relative to each) with one full rotation. For example, an M8 screw has a standard pitch of 1.25mm. So if you have a lead screw with an M8 thread, every full rotation of the lead screw will move the nut 1.25mm.

polycarbonate A very strong plastic that can be 3D printed. This is the type of plastic that bulletproof windows are made from.

polyimide film Often referred to by its trade name Kapton, a film that can be used as a 3D printing surface to improve adhesion. It can also be used as a tape in very high-temperature applications, such as on the hot end.

powder bed printing Any type of 3D printing process that applies a liquid binder to powder.

print fan A small fan that cools the extruded plastic after it has been deposited.

rapid prototyping A term that is often used synonymously with 3D printing.

RepRap This can refer to the RepRap project, which develops open-source 3D printers, as well as the 3D printers themselves.

reverse engineering The process of determining the function and/or design of a man-made object or system. This can be anything from reverse engineering software to complicated mechanical systems.

scale The size of a part in relation to its physical real-world counterpart. It is often used to represent something large (like a building) at a manageable size.

selective laser sintering (SLS) A 3D printing process that uses a laser to essentially melt a powder into a solid piece. This process can be used for 3D printing metal parts.

shield In the context of Arduinos, a circuit board designed to be attached to the Arduino board. Shields are generally used to expand the capabilities of Arduinos by adding either more connections or sensors. They are usually designed for a specific application (like GPS tracking or to interface with another device), although plenty of general-use shields exist.

slicing software   The 3D printing software that takes a 3D model and converts it into a series of instructions for the 3D printer.

smooth rod   A cylindrical metal rod used in 3D printers for linear motion.

square-cube law   Describes the mathematical relationship between area and volume. Its relevance to 3D printing is the way that volume (and therefore print time) increases exponentially with size. For example, a cube 10mm to a side has a volume of 1,000mm³. Doubling the dimensions of the cube to 20mm to a side results in a volume of 8,000mm³. Because the larger cube is 8 times the volume, it will take 8 times longer to print than the smaller cube.

stepper motor   A type of electric motor commonly used for 3D printers because its rotation can be very precisely controlled.

stereolithography   The very first 3D printing process, this uses a focused UV laser to cure photopolymer resin.

STL   A file format used to store 3D models. This file format is used by almost all 3D printing software, as well as CAM software.

surface finish   The quality (generally smoothness) of the surface of a printed part.

tangent   Refers to a line that touches a curve at a single point and continues straight on from that point. The easiest way to visualize this is to picture a circle. A tangent line touching the far-right side of the circle will continue on vertically (up, down, or both), touching just a single point of the circle. The actual mathematics of defining a tangent line are fairly complicated, but luckily CAD software handles all of that—you only need to be concerned with the practical effect.

thermistor   An electronic component that changes resistance based on heat. It's usually used to monitor the temperature of the hot end and heated bed.

thermocouple   A type of temperature-measuring device used in a wide range of industries. It's inexpensive and doesn't require a power source, which makes it ideal for some applications. However, thermocouples aren't very accurate, which generally makes them unsuitable for use in 3D printer hot ends.

threaded rod   A type of fastener that is basically a long bolt without a head at the end. A threaded rod is often used in place of lead screws on 3D printers.

units   The unit of measurement used in 3D modeling and printing. This specifies what is used for the coordinates—for instance, millimeters or inches.

Z height   The distance between the tip of the nozzle and the print bed. This is adjustable and is important for adhesion.

APPENDIX B

Resources

If you've finished reading this book and you're looking for more material, or if you just need some help for your specific 3D printer, this is the place to look. The following are some online community resources I've compiled that may be helpful to you.

General Resources

These are online resources that should be applicable to just about every consumer 3D printer. These websites can give you information on troubleshooting, printer modifications, and printing in general.

3D Printing Subreddit (reddit.com/r/3Dprinting/): If you're familiar with Reddit, be sure to check out this subreddit. It's the most active 3D printing subreddit and has many knowledgeable users.

RepRap (reprap.org): This community-edited resource is full of information on RepRap and RepRap-derived 3D printers. It also contains a lot helpful information on various 3D printer components that are used in all consumer 3D printers.

RepRap Forums (forums.reprap.org): This is another RepRap community, but because it's a forum, you can ask questions and get help from other forum members. Before asking questions, be sure to use the search function to see if your questions have already been answered.

RepRap Subreddit (reddit.com/r/reprap/): Another Reddit community, this one is specifically for discussions about RepRap 3D printers (and their derivatives).

Printer-Specific Resources

This is a list of websites dedicated to specific 3D printer models or brands. If you need in-depth information on your particular 3D printer, these are some websites to check.

Deezmaker Bukobot Forums (forum.bukobot.com)

Flash Forge Knowledge Base (http://www.flashforge-usa.com/support/)

LulzBot Forums (forum.lulzbot.com)

MakerBot Community Links (makerbot.com/blog/)

MakerBot Subreddit (reddit.com/r/makerbot/)

Printrbot Subreddit (reddit.com/r/printrbot/)

Printrbot Talk Forums (printrbottalk.com)

SeeMeCNC Forums (forum.seemecnc.com)

Solidoodle Forums (soliforum.com)

Solidoodle Subreddit (reddit.com/r/solidoodle/)

Ultimaker Forums (umforum.ultimaker.com)

Ultimaker Subreddit (reddit.com/r/ultimaker/)

Miscellaneous Resources

Here you'll find a short list of websites related to 3D printing, but not directly. These will be good for further reading and research for your 3D printing projects.

Instructables (instructables.com): This website has detailed instructions for many kinds of DIY projects. A lot of these take advantage of 3D printing, and this can be a good place to find ideas for 3D printing projects.

Make: (makezine.com): The online website for *Make:* magazine, which is a publication for makers and tinkerers. The same company organizes Maker Faires, which are great local events with lots of 3D printing booths and displays.

MakerBot Thingiverse (thingiverse.com): This is, by far, the largest and most active 3D model repository that is specifically for 3D printing. As a user, you can download 3D models, free of charge, to print on any 3D printer. You can also upload your own models to share with others.

Serial Hobbyism (serialhobbyism.com): This is my personal website, where I post articles on various hobbies (including 3D printing). If you still want to read more from me after this book, be sure to check it out!

APPENDIX C

Further Uses of 3D Printing

You've read through this entire book and even managed to reach the final appendix! Obviously, you just can't get enough of 3D printing. Luckily, there are a lot of uses for 3D printing I haven't gone into yet. Some of these require more than just a regular 3D printer, while others are very new and experimental ways of using a 3D printer. Whatever the case, most of these will also require additional research on your part. But this should give you an idea of some more things you can do with your 3D printer now and potentially in the future.

Your 3D Printer as an Intermediary Process Tool

One way to take advantage of your 3D printer is to use it as an intermediary for other creative processes. Think of it like how you use a mixer before using the oven when you bake a cake. The 3D printer becomes just one tool in the process instead of doing everything. Using this approach, you can get around the limitations of 3D printing, especially when it comes to material options and producing large quantities of parts.

Investment Casting

Investment casting is an extremely old manufacturing technique. It's the process of making a part, surrounding it with a mold, and then pouring molten metal into the mold to reproduce the part. The reason it's called *investment casting* is because the original part is lost during the process and the mold cannot be reused.

Traditionally, investment casting was done with a wax original part. The wax was easy to work with, melted easily, and was cheap, allowing a person to make an intricate metal part by first working with wax. Investment casting was (and is) very popular for making jewelry.

But astonishingly, 3D printers are almost perfect for this process as well. You can design a part in CAD, 3D print it in PLA, and use that printed part for investment casting. The melting point of PLA is low enough that it's suitable for investment casting.

With this process, you don't need a wildly expensive SLS 3D printer to make metal parts. Instead, you can just use your inexpensive hobby printer to print PLA parts, and then use investment casting to make metal parts. This is a method that has been proven to work very well and is actually relatively inexpensive.

So what do you need to actually do this? Well, first you need a way to melt the metal, which is a job handled by a foundry. Aluminum foundries can be made at home for less than $100. You'll also need a handful of basic tools and safety equipment for handling the molten metal. And finally, you'll need some casting sand for making the molds themselves.

All in all, you can pretty easily get set up for investment casting for just a couple hundred dollars. Home metal foundries are usually only capable of handling aluminum (which has a relatively low melting point), but that's good enough for most hobbyists.

A whole book could be written on investment casting alone, but the basics are easy enough to understand:

- Once you've printed a part in PLA, fill a container about halfway full with your casting sand. There are many formulas in use for the sand, but it's generally normal sand mixed with some sort of binder, like plaster.

- Push half the part into the sand, making sure all of the crevices of the part are filled in.

- Fill the rest of the container with sand, with a sprue (which forms a channel) for pouring the metal and another to vent the air and gases as the metal is poured.

- Once the casting sand has had time to set (the time frame will vary dramatically depending on the sand formula), remove the hardened block of sand from the forming container.

- Using the foundry, melt the metal (usually aluminum for hobby casting) in a crucible. Once it's molten, it can be poured into the opening formed by the sprue. As it's poured in, the molten metal will melt away the original PLA part.

- After the metal has cooled, the block of sand can be broken apart, and the metal part can be removed for finishing.

It may take some practice, but soon you'll be able to make your own 3D metal parts at home!

Mold Making and Resin Casting

3D printing is great for making a few parts, but what if you want to make many copies of the same part? Investment casting destroys the original part and the mold, so it's not a good option if you want multiple copies. This is where mold making and resin casting come in. They allow you to make many copies of a part by simply pouring molds. This is a more time- and cost-effective method of creating duplicate parts than simply 3D printing them.

The first way to go about this is to 3D print the mold itself (which is the negative). This will require you to either use mold-making tools in CAD or manually model the mold in CAD. With a model of the mold, you can simply 3D print it like any other part.

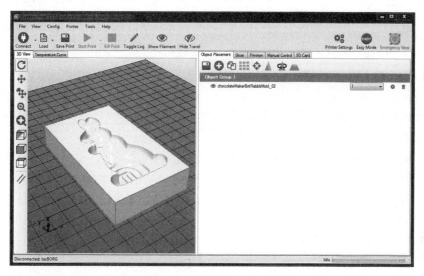

A one-piece mold for making a chocolate bunny.

Once you have your mold, you can use common hobby casting materials to make your parts. However, not all materials will be suitable, because some will stick to the mold. But many materials (such as silicone) can be used.

All you have to do is pour the liquid material into the mold and wait for it to set and cure. Depending on the material, this could take anywhere from a few minutes to a couple of days. Once it's cured, you can just pull it out of the mold and start a new part.

The other way of casting adds an extra step but has a number of benefits. For this process (often called *resin casting*), you print a normal part on your 3D printer exactly as you normally would. You then use a flexible silicone mold-making rubber to make a mold around the printed part (the positive). After the silicone has cured, you can remove the original part and begin casting new parts by pouring resin into the silicone mold.

The benefits of the resin casting method are twofold: you can make many molds from a single printed part which can be used simultaneously, and your final parts can be made from a variety of materials. With the resin casting method, you can make your final parts from materials like polyurethane, which is incredibly strong—and much stronger than the original printed part.

Using Your Printer for Good

But maybe you're not just looking for another way of making parts. Maybe you want to do a little more, and maybe you're even feeling philanthropic. As luck would have it, 3D printing can help with that, too!

Charity Work

One of the most exciting things 3D printers are being used for right now is charity work. One use in particular has grabbed the attention of many people: 3D printing prosthetics.

Traditionally, prosthetics have been very utilitarian and rather unwieldy. They're expensive and tend to be ill-fitting and difficult to use. Custom prosthetics have always been much better, but they're even more expensive and aren't usually covered by medical insurance.

That's where 3D printing has come in. Under the direction of organizations like e-NABLE (enablingthefuture.org), 3D printer owners have been able to give a hand to those in need—literally. People with access to 3D printers can 3D print prosthetic hands and give them to the people who need them.

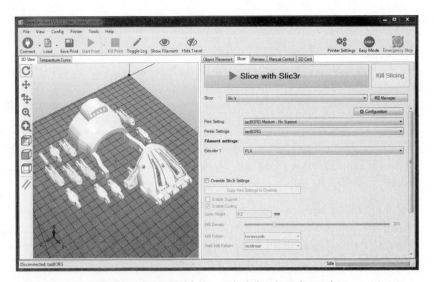

All of the parts needed for a prosthetic hand, ready to print.

I can't overstate how revolutionary this is. The prosthetics themselves are rapidly becoming more and more advanced as they're developed further. They can be customized to fit the user perfectly. And, perhaps best of all, they're very inexpensive to 3D print.

This has allowed those who have lost limbs to get high-quality prosthetics for free. And those prosthetics aren't junk either; in most cases, they're much better than the prosthetics they would have received through traditional avenues.

Perhaps most importantly, the e-NABLE project is hugely beneficial to children. Because children grow so quickly, it's difficult for them to get prosthetics that fit them as they grow. But now they can receive new 3D printed prosthetics that fit them perfectly as they get older.

So if you have any interest in giving back, I highly recommend you look into this project (and others). With just a little bit of your time and few bucks worth of filament, you can really make a difference in someone's life.

Hackerspaces

If the idea of getting involved in making things with other people appeals to you, consider joining a local hackerspace (hackerspaces.org) or makerspace in your area. Hackerspaces are basically small community workspaces where members can use a variety of tools for their own personal projects. The idea is pretty simple: tools are expensive, so why not share them?

Most big cities in the United States have one or more hackerspaces, and it's likely there is one in your area. To join, you usually pay a monthly fee to gain access to the hackerspace. Once you're a member, you'll have access to a wide range of tools like 3D printers, CNC mills, laser cutters, welders, carpentry tools, and so on.

Additionally, this will give you a place to congregate with other like-minded people. You can collaborate on projects, get help with challenging problems, and take classes to expand your skill set. Some hackerspaces are focused on particular topics (like electronics or metal working), while others are more general and have a little bit of everything.

If you do join a hackerspace and have a 3D printer, it's certainly worth considering bringing your 3D printer. They do use membership dues to purchase tools, but hackerspaces are expensive to run. A lot of the tools there are owned by members or were donated by members. Whether you decided to bring your 3D printer or not, hackerspaces and makerspaces are a great way to get involved with people in your community.

Potential Applications for 3D Printing

So far, I've talked about things you can use your 3D printer for. But maybe you're interested in learning about some experimental uses of 3D printing currently in development. If so, there are many interesting applications for 3D printing that are currently being explored.

For now, these uses of 3D printing are still experimental and under development, but the possibilities for improving people's lives are very interesting. They're still very new, but as these technologies mature, you'll start to see how they can truly affect people's lives around the world.

Printing Food

FFF is unique among 3D printing technologies, because the same basic concept that allows an FFF printer to extruder molten plastic can also work for any other material that can be liquefied and then solidified. If a substance can be liquefied, squeezed out of a nozzle, and then solidified, it can theoretically be 3D printed using the FFF printing process. That can even include food.

Why would anyone want to 3D print food? Right now, my impression is that it's mostly a matter of novelty. But there are some reasons why it could be more than just a curiosity. One example of this is wedding cakes. Anyone who has ever had to shop for wedding cakes knows that intricate detail in the decoration is highly sought after. Those decorations are carefully crafted by hand by the baker, which can be difficult and time-consuming, which means it's expensive.

But what if those intricate decorations could be 3D printed? It's possible to use an FFF 3D printer to extrude icing onto a cake and to create complex decorations with the icing. In fact, it's already been done. Icing is stored in a tube which can be compressed to squeeze the icing through a nozzle. Just like with a standard FFF printer, the icing can then be layered to create 3D designs.

Cake decorations are just one example of 3D printing food, and admittedly it's not likely to change the world. But what if 3D printing could be used in an unconventional way that does have the potential to change the world?

Creating Buildings

If you've ever seen a new building being constructed, you know it can take weeks or months. First, a foundation needs to be poured. After that, a frame has to be constructed, walls must be added, a roof needs to be built, siding has to be added, and so on. It's a lengthy and complicated process, and a lot of labor is needed for the construction. But what if buildings themselves could be 3D printed?

That's not only possible, but it's already been done. It's still very experimental, of course, and is mostly being done to test the idea. But it is being done, and the idea has been proven. The process works by extruding concrete, layer by layer, to form the structure of the house.

The idea of 3D printing houses or other buildings has tremendous potential for changing the world. For one, automating the process of construction means the labor needed is greatly reduced. Also, a large-scale 3D printer can also run all day and night, reducing the time it takes to construct a building.

The possibilities are almost limitless for this technology. Strong and durable homes can be built quickly and cheaply without skilled labor. By mounting the concrete extruder on a crane, the potential size of the 3D printed building is almost completely unrestricted. Building plans can be easily reused, reducing the time and cost of designing them.

You can probably easily imagine the implications in third-world and developing countries. Long-lasting homes and buildings could be constructed quickly and efficiently, making a big difference in countries where families can't afford homes constructed using traditional means.

The potential for improving the living conditions of people around the world is fantastic. But what if 3D printing could go even further? What if it could actually save lives?

Making Organs

Even the wealthiest countries in the world struggle with providing adequate health care to their citizens. One complication in particular is holding back medical care: organ transplants. Two problems make organ transplants especially difficult: finding organ donors and rejection of the transplanted organ.

Finding an organ to transplant is problematic for obvious reasons: it requires a donor. For some organs, like kidneys, the donor can be alive, but most people aren't eager to part with their organs. For others—like hearts, for example—the donor has to be deceased. In both cases, it's difficult to find a donor. Even if a donor is found, blood types have to match between the donor and the recipient of the transplant (among other requirements).

After all of that has been done, it's still possible for a transplanted organ to be rejected by the recipient. The human body just isn't too accepting of tissue that originated in another body.

Luckily, researchers think 3D printing can solve both of these problems. The idea is that tissue is harvested from the patient's own body and used as the material to 3D print a new organ. This means a donor isn't needed, and also means that the patient's body is likely to accept the new organ. This possibility alone is enough to revolutionize medicine, but it doesn't stop there.

When pharmaceutical companies test new drugs or treatments, they need something to test them on. Traditionally, this is done on animals or human patients, which can be risky for both. But if living tissue can be 3D printed, it would be possible to ethically test drugs without the risk of harming any humans or animals. It also means tissue can be printed with a specific condition that the drug is designed to treat.

For example, if a pharmaceutical company wanted to develop a new treatment for cancer, they could 3D print tissue with cancer. That would let them test the treatment on real human tissue that has cancer without risking the well-being of an actual cancer patient. Done on a large scale, this could dramatically reduce the time it takes to develop a new drug while simultaneously increasing the effectiveness of that drug.

Index

A

ABS (acrylonitrile butadiene styrene), 44
    build platform surface treatment, 98
acetone, 98
additive manufacturing, 24-27
adhesion problems, troubleshooting, 169-171
adjusting Z height, 145-146
Aleph Objects, 6-7
algorithms, slicers, 131
all-metal hot ends
    filament materials, 289
    switching to, 300
arc command (CAD), 222
Arduino, 16-17
    control boards, 106-107
assembled 3D printers, 116-117
assemblies, CAD, 231
    planning, 232-233
attachment components, 70-73
Autodesk 123D CAD software, 217
Autodesk Inventor CAD software, 217
auto-leveling beds, 142-143, 171
automatically adjusting Z height, 144-145
axes, extending, 302
axis command (CAD), 221

B

backlash, 52
    anti-backlash nut design, 72-73
Bakelite, 290
ball bearings, 62-63
Basic Input/Output System (BIOS), 128

Beaman, Joe, 15
bearings, 62-63
beds
    See also build platforms
    auto-leveling beds, 142-143, 171
    filament materials, 290-291
    heated
        adding, 297
        build platforms, 92-94
        filament materials, 290-291
        preheating, 175-176
        size, 297-298
    leveling, 140
        auto, 142-143
        manually, 140-142
    size, 297-299
Bed size setting (slicing software), 150
belts, 65-66
BIOS (Basic Input/Output System), 128
blobs, troubleshooting, 171
borosilicate, 92
Bowden cables, 78-80
    direct drive, 80-81
    geared, 80-81
Bowden, Ernest Monnington, 79
Bowden, Frank, 79
Bowyer, Adrian, 16
Bridges fan speed setting (slicing software), 154
Brim width setting (slicing software), 156
build plate. *See* build platform
build platform, 91
    heated beds, 92-94
    heated build chambers/enclosures, 94
    materials, 92

surface treatments, 95
  ABS juice, 98
  hairspray, 98-99
  painter's tape, 95-96
  PET (polyethylene terephthalate), 97
  polyimide film, 96-97
  white glue, 96
bushings, 62-63

C

CAD (computer-aided design) software, 4-5, 14, 213, 227
  choosing, 217
  commands, 218
    modeling, 218-222
    sketching, 222-223
  modeling, 228-230
    assemblies, 231-233
    exporting files, 236-237
    fitting parts, 231-235
    successful 3D parts, 235-236
  origins, 214
  pattern tools, 229
  premodeling, 228
  scaling, 225-226
  simulation add-ons, 214
  units, 223-224
  versus 3D modeling, 215-217
calipers, 153, 241
CAM software, subtracting material, CNC mills, 28-29
Carabiner project, 175
  extruding filament, 179-180
  letting cool, 182-183
  loading filament, 178-179
  loading .STL file, 176-177
  monitoring first layer, 181-182
  preheat extruder and heated bed, 175-176
  removing, 183-184
  slicing model, 177-178
  starting print, 181
Cartesian, 50
Cartesian 3D printers, 50, 122
  construction, 50-51
  layout, 51-53
ceramics, printing, 9
chamfer, 222
chamfer command (CAD), 222
circle command (CAD), 222
circles, extruding, 252, 253
closed-source hardware, 116
CNC (computer numerical control) mills, 23, 27-29
  converting to PCB mills, 302-303
  end stops, 102
  milling materials, 29
  spindles, 27
  subtracting material with CAM software, 28
  subtractive manufacturing, 28-29
  versus 3D printers, 23-24, 29, 34
    cost, 29-30
    materials, 32
    part geometry, 30-31
    surface finish, 32-34
cold ends
  Bowden cables, 78-80
  direct drive, 80-81
  direct feed, 77-78
  (extruders), 76-81
  geared, 80-81
commands
  CAD, 218
    modeling, 218-222
    sketching, 222-223
  extrude, 134
  fan power, 134
  G20 (units to inches), 137
  G21 (units to mm), 137
  G28 (homing), 137
  G29 (auto-leveling), 137
  G92 (setting coordinates), 137
  heated bed power, 134
  home all, 134
  home axis, 134

hot end power, 134
jog axis, 133
M92 (setting steps), 137
M119 (end stop status), 137
M500 (storing EEPROM values), 137
M503 (reading EEPROM values), 138
retract, 134
stopping motor power, 134
COM port, choosing, 132
components
   attachment and connection, 70-73
   movement, 59-60
      bearings, 62-63
      belts, 65-66
      pulley systems, 65-66
      rails, 61-62
      smooth rods, 61-62
      stepper motors, 63-65
   weight-bearing, 67
      lead screws, 68-69
      threaded rods, 69-70
computer-aided design (CAD). *See* CAD (computer-aided design) software
computer numerical control, 27
computer numerical control (CNC) mills. *See* CNC (computer numerical control) mills
configuring slicing software, 148
   filament settings, 152-154
   printer settings, 150-152
   print settings, 154-157
connections, 70-73
   control board, 132
   frame component, 55
   host software, 132-133
constraints command (CAD), 223
construction
   3D printer frames, 53-56
   Cartesian 3D printers, 50-51
continuous-rotation servos, 64
control board, 106
   Arduino, 106-107
   changing, 108-109
   connecting to computer, 132
   G-code, 135-138
   installing drivers, 132-133
   proprietary, 106-107
   SD card support, 110
   upgrading, 108-109, 298
control components, 101
   control boards, 106-109
   end stops, 102-105
   LCD controllers, 110-112
   SD cards, 109-110
controlling 3D printers, 133-134
cooling
   cracking, troubleshooting, 168
   printing filament materials, 292-293
correct Z height, knowing, 145-146
cost
   3D printers, 29-30, 121-122
   CNC mills, 29-30
couplers, 70-72
cracking
   FFF printers, 44-45
   troubleshooting, 168
Creo CAD software, 217
Crump, Lisa, 15
Crump, Scott, 15
Custom Storage Drawer project, 259
   adding fillets to handle, 264
   adding ridges for grip, 264-265
   adding rough handle, 263
   creating new part, 259
   cutting opening for handle, 260-261
   designing compartments, 261-262
   exporting .STL file, 265
   extruding body of drawer, 260
   printing, 265

D

DARPA (Defense Advanced Research Projects Agency), 15
Darwin, RepRap, 17-19

Deas, Nathan, 193
Deckard, Carl, 15
Defense Advanced Research Projects Agency (DARPA), 15
Delta-style 3D printer, 122
Descartes, René, 50
design
    extruder hot ends, 83-85
    frame construction, 54
design intention, reverse engineering, 245-247
Diameter setting (slicing software), 153
digital calipers, 241
digital light processing (DLP) printing, 38-39
dimensional accuracy, troubleshooting, 165-166
dimension command (CAD), 223
direct drive cold ends, 80-81
direct feed cold ends, 77-78
    direct drive, 80-81
    geared, 80-81
direct metal laser sintering (DMLS), 42
Disable fan (slicing software), 154
discoloration, troubleshooting, 169
Distance from object setting (slicing software), 156
DIY (do-it-yourself) 3D printers, building, 117-118
DLP (digital light processing) printing, 38-39, 123
DMLS (direct metal laser sintering), 42
downloading model pack, 175
drafting, 215
drivers, installing on control board, 132-133
drooping, troubleshooting, 172-173
Dust Collector project, 267
    copying clip, 271-273
    creating new part, 267-268
    creating second part, 269-270
    cutting groove, 268-269
    exporting .STL file, 273-274
    making first clip, 270-271
    printing, 273-274
    revolving body, 267-270

E

eBay, 304
ellipse command (CAD), 222
Elmer's glue, build platform surface treatment, 96
Enable auto cooling setting (slicing software), 153
Enable fan if layer print time is below setting (slicing software), 154
End G-code setting (slicing software), 151
end stops, 102, 105
    CNC mills, 102
    mechanical, 102-104
    optical, 104-105
    switches, 102-104
engineering
    CAD, 227
        commands, 218-223
        exporting files, 236-237
        modeling, 228-236
        options, 217
        origins, 214
        premodeling, 228
        scaling, 225-226
        units, 223-224
        versus 3D modeling, 215-217
    reverse, 239-240
        benefits, 240
        inferences, 244-248
        measurement tools, 240-242
        measuring parts, 243
        modeling parts, 248-249
exporting CAD files, 236-237
extend command (CAD), 223
Extra length on restart setting (slicing software), 152
extrude command (CAD), 134, 218
extruded cut command (CAD), 218
extruders, 75-76
    3D printers, 31
    cold ends, 76-81
        Bowden cables, 78-80
        direct drive, 80-81
        direct feed, 77-78
        geared, 80-81

hot ends, 76-85
    compatibility, 84
    heating elements, 83
    physical design, 83-85
    thermistors, 82-83
installing multiple, 301-302
jammed, troubleshooting, 164-165
multiple, 87-88
nozzles, 85-86
preheating, 175-176
print fans, 86-87
Extruders setting (slicing software), 151
extruding, 76
    circles, 252-253
    filament, 26-27, 179-180
    troubleshooting problems, 163-166
Extrusion multiplier setting (slicing software), 153

F

failed prints, reusing, 162
fan power command, 134
fan shrouds, adding, 296-297
Fan speed setting (slicing software), 154
FDM (fused deposition modeling) printers, 9, 15
    extruders, 75
        cold ends, 76-81
        hot ends, 81-85
        multiple, 87-88
        nozzles, 85-86
        print fans, 86-87
FFF (fused filament fabrication) printers, 9, 15, 25, 37, 42-45
    cost, 43
    cracking, 44
    print times, 44
    quality, 43
    simplicity, 43
    warping, 44-45
filament materials, 42
    different properties, 88
    extruding, 26-27, 179-180

flexible, 88, 120-121, 286-287
hardware requirements, 288-291
high-impact polystyrene (HIPS), 288
loading, 178-179
nylon, 286
polycarbonate, 286
polyethylene terephthalate (PET), 288
printing techniques, 291-293
wood, 287
filament settings, slicing software, 152-154
files, exporting CAD files, 236-237
Fill density setting (slicing software), 156
Fill pattern setting (slicing software), 156
fillet command (CAD), 220-222
Fillet tool, 254
finishes
    CNC mills versus 3D printers, 32-34
    surface, 11-12
First layer bed temperature setting (slicing software), 153
First layer extruder temperature setting (slicing software), 153
First layer height setting (slicing software), 155
firmware, 128-129
flexible filament, 88, 120-121, 286-287
frames (printers), 49
    Cartesian, 50
        construction, 50-51
        layout, 51-53
    component connections, 55
    construction, 53-56
    design, 54
    material, 55-56
    size, 56-57
    t-slot aluminum extrusion, 55-56
FreeCAD, 217
Free Software Foundation (FSF), 114
fused deposition modeling (FDM) printers. *See* FDM (fused deposition modeling) printers
fused filament fabrication (FFF) printers. *See* FFF (fused filament fabrication) printers

G

G20 (units to inches) command, 137
G21 (units to mm) command, 137
G28 (homing) command, 137
G29 (auto-leveling) command, 137
G92 (setting coordinates) command, 137
Garolite, 290
G-code, 131-135, 150
    manual control functions, 136-138
    printing, 135-136
G-code flavor setting (slicing software), 150
geared cold ends, 80-81
geometric inferences, reverse engineering, 244
geometry
    overhangs, 11
    parts, 30-31
ghosting, 80
    troubleshooting, 173
green sand, 42

H

hairspray, build platform surface treatment, 98-99
Hall, Edwin, 105
Hall effect, 105
hardware requirements, filament materials, 288
    all-metal hot ends, 289
    beds, 290-291
    heated beds, 290-291
    print fans, 289
    printing techniques, 291-293
heated bed power command, 134
heated beds
    adding, 297
    build platforms, 92-94
    filament materials, 290-291
    preheating, 175-176
    size, 297-299
heated build chambers/enclosures, build platform, 94
heating elements, 83

helix/spiral command (CAD), 221
HIPS (high-impact polystyrene), 88, 288
hole command (CAD), 220
home all command, 134
home axis command, 134
host software, 129-130
    connecting, 132-133
    manual control functions, 133-134
    versus slicing software, 131
hot end power command, 134
hot ends (extruders), 26, 76, 81-85
    all-metal, 289, 300
    compatibility, 84
    heating elements, 83
    physical design, 83-85
    thermistors, 82-83
    troubleshooting temperature, 166-168
Hull, Charles W., 14, 15, 38
hypotenuse, 244

I-J

inferences, reverse engineering, 244
    design intention, 245-247
    geometric, 244
    proportional, 247-248
injection molding, 6
intellectual property, 114

jammed extruders, troubleshooting, 164-165
jog axis command, 133

K-L

Kapton tape, 96, 167
Keep fan always on setting (slicing software), 153
kits, 3D printers, 117

laser cutting, alterations, 303-304
Layer change G-code setting (slicing software), 151
Layer height setting (slicing software), 155
layer ridges, 11-12

layer thickness, printing filament materials, 293
layers, 25
    filament extrusion, 26-27
    slicing and creating for, 24-26
layout, Cartesian 3D printers, 51-53
LCD controllers, 110-112
lead screws, 68-69
leveling beds, 140-143
Lift Z setting (slicing software), 151
limonene, 88
line command (CAD), 222
linear bearings, 63
loft command (CAD), 219
lofted cut command (CAD), 220

M

M92 (setting steps) command, 137
M119 (end stop status) command, 137
M500 (storing EEPROM values) command, 137
M503 (reading EEPROM values) command, 138
MakerBot, 10, 94
maker culture, 8
manual control functions, G-code commands, 136-138
manual control functions, host software, 133-134
manually adjusting Z height, 144-145
manually leveling beds, 140-142
manufacturing, 6-7
    additive process, 24-27
    CNC mills, subtractive process, 28-29
    filament extrusion, 26-27
    slicing and creating layers for models, 24-26
materials, 9, 55-56, 285
    build platform, 92
    CNC mills versus 3D printers, 32
    flexible filament, 286-287
    green sand, 42
    hardware requirements, 288-291
    high-impact polystyrene (HIPS), 288
    milling, 29
    nylon, 286
    polycarbonate, 286
    polyethylene terephthalate (PET), 288
    printing techniques, 291-293
    wood filament, 287
McMaster-Carr, 304
mechanical advantage, 65
mechanical end stops, 102-104
Mendel, RepRap, 18
metals, printing, 9
microstepping, stepper motors, 64-65
Min print speed setting (slicing software), 154
Minimum extrusion length setting (slicing software), 156
Minimum travel after retraction setting (slicing software), 152
mirror command (CAD), 221-223
MJP (MultiJet Printing), 41
model pack, downloading, 175
modeling
    CAD (computer-aided design), 228-230
        assemblies, 231-233
        exporting files, 236-237
        fitting parts, 231-235
        premodeling, 228
        successful 3D parts, 235-236
    parts, reverse engineering, 248-249
    versus CAD (computer-aided design), 215-217
modeling commands (CAD), 218, 219, 220, 221, 222
models
    filament extrusion, 26-27
    resizing, 185-186
    slicing, 177-178
    slicing and creating layers, 24-26
Monogrammed Coaster project, 251
    creating new part, 251
    cutting letters, 254-256
    exporting .STL file, 256-257
    extruding circles, 252-253
    filleting top edge, 254
    printing, 256-257
motors, stepper, 63-65

movement components (3D printers), 59-60
　bearings, 62-63
　belts, 65-66
　pulley systems, 65-66
　rails, 61-62
　smooth rods, 61-62
　stepper motors, 63-65
MultiJet Printing (MJP), 41
multiple extruders, 87-88
　installing multiple, 301-302
Multiple extruders setting (slicing software), 156

N-O

Nozzle diameter setting (slicing software), 151
nozzles (extruders), 85-86
　compatibility, 86
numerical control, 27
nuts, 72-73
nylon, 286

offset command (CAD), 223
OpenSCAD, 217
Open Software Initiative (OSI), 114
open source, 3D printers, 114-116
open source, RepRap project, 16-17
optical end stops, 104-105
Other layers bed temperature setting (slicing software), 153
Other layers extruder temperature setting (slicing software), 153
overextrusion, troubleshooting, 163
overhangs, 11

P-Q

painter's tape, build platform surface treatment, 95-96
parts
　availability, 20-21
　CAD (computer-aided design), 231-236
　finding, 304-305
　geometry, 30-31
　price, 21-22
　reverse engineering, 239-240
　　inferences, 244-248
　　measurement tools, 240-242
　　measuring parts, 243
　　modeling, 248-249
pattern command (CAD), 223
pattern tools, CAD, 229
PCB mills, 302-303
PCBs (printed circuit boards), 92-93
Pencil Holder project, 185
　loading filament, 186-187
　loading .STL file, 185-186
　monitoring first layer, 189-190
　preheating extruder and heated bed, 186-187
　removing, 190-191
　resizing model, 185-186
　slicer settings, 187-188
　slicing models, 188
　starting print, 189
Perimeters (minimum) setting (slicing software), 155
PET (polyethylene terephthalate), 288
　build platform surface treatment, 97
phenolic resin plastic, 290
photo interrupter, 104
photopolymer resin, 14
pitches, lead screws, 68
plane command (CAD), 221
point command (CAD), 221
polycarbonates, filaments, 121, 286
polyethylene terephthalate (PET), 288
polygon command (CAD), 222
polyimide film, build platform surface treatment, 96-97
PolyJet Printing, 41
polyvinyl acetate (PVA)-based white glue, build platform surface treatment, 96
Position setting (slicing software), 151
powder bed printing, 39-40
power supplies, upgrading, 298-299
premodeling, CAD, 228

price
   3D printers, 21-22, 29-30, 121-122
   CNC mills, 29-30
print beds
   See also build platforms
   auto-leveling beds, 142-143, 171
   filament materials, 290-291
   heated
      adding, 297
      build platforms, 92-94
      filament materials, 290-291
      preheating, 175-176
      size, 297-298
   leveling, 140
      auto, 142-143
      manually, 140-142
   size, 297-299
Print center setting (slicing software), 150
print fans (extruders), 86-87
   filament materials, 289
print settings, slicing software, 154, 155, 156, 157
print volume, 119
printed circuit boards (PCBs), 92-93
printer settings, slicing software, 150-152
printers
   See also 3D printers
   MJP (MultiJet Printing), 41
   RepRap, rapid development, 17-18
printing
   filament materials, 291
      cooling, 292-293
      layer thickness, 293
      speed, 292
      temperature, 291-292
   G-code, 135-136
   preparation, 157
   running prints, 158-159
   time, 11
prints
   resolution, 119-120
   reusing failed, 162
   starting, 181
   troubleshooting, 161-162, 171-173
      adhesion problems, 169-171
      extrusion problems, 163-166
      temperature problems, 166-169
projects
   Carabiner, 175
      extruding filament, 179-180
      letting cool, 182-183
      loading filament, 178-179
      loading .STL file, 176-177
      monitoring first layer, 181-182
      preheat extruder and heated bed, 175-176
      removing, 183-184
      slicing model, 177-178
      starting print, 181
   Custom Storage Drawer, 259
      adding fillets to handle, 264
      adding ridges for grip, 264-265
      adding rough handle, 263
      creating new part, 259
      cutting opening for handle, 260-261
      designing compartments, 261-262
      exporting .STL file, 265
      extruding body of drawer, 260
      printing, 265
   Dust Collector, 267
      copying clip, 271-273
      creating new part, 267-268
      creating second part, 269-270
      cutting groove, 268-269
      exporting .STL file, 273-274
      making first clip, 270-271
      printing, 273-274
      revolving body, 267-270
   Monogrammed Coaster, 251
      creating new part, 251
      cutting letters, 254-256
      exporting .STL file, 256-257
      extruding circles, 252-253
      filleting top edge, 254
      printing, 256-257

Pencil Holder, 185
    loading filament, 186-187
    loading .STL file, 185-186
    monitoring first layer, 189-190
    preheating extruder and heated bed, 186-187
    removing, 190-191
    resizing model, 185-186
    slicer settings, 187-188
    slicing models, 188
    starting print, 189
Reverse Engineering, 275-276
    adding screw hole, 281-282
    creating new part, 276
    creating screw hole support, 280-281
    cutting socket opening, 278-279
    exporting .STL file, 282
    extruding body, 276
    filleting edges, 276-277
    mirroring socket opening, 279
    printing, 282
    shelling cover, 277-278
Robot, 193
    loading filament, 194
    loading .STL file, 193
    modifying slicer settings, 194-196
    monitoring first layer, 197
    preheating extruder and heated bed, 194
    printing feet down, 193
    removing, 198-199
    removing supports, 200-202
    slicing model, 196
    starting print, 197
Storage Box, 203
    loading drawer .STL file, 207-208
    loading storage body .STL file, 204-205
    printing drawer, 208-210
    printing storage body, 206-207
    slicing storage body, 205-206
Pronterface, 133
proportional inferences, reverse engineering, 247-248
proprietary control boards, 106-107
prototyping, rapid, 4-6
Prusa Mendel, RepRap, 18
pulley system, planning assembly, 232-233
pulley systems, 65-66
PWM (pulse-width modulation), 84
Pyrex, 92

R

Raft setting (slicing software), 156
rails, 61-62
RAMPS control board, 16
rapid prototyping, 4-6
rectangle command (CAD), 222
Repetier host software, 130
replicators, 10
RepRap Arduino Mega Pololu Shield (RAMPS) control board, 16
RepRap printers, 16-17
    couplers, 71
    Darwin, 17-19
    Mendel, 18
    open source, 16-17
    Prusa Mendel, 18
    rapid development, 17-18
resistive heating, 83
retract command, 134
retraction, 171
Retraction length setting (slicing software), 151
Retract on layer change setting (slicing software), 152
reusing failed prints, 162
reverse engineering, 239-240
    benefits, 240
    inferences, 244
        design intention, 245-247
        geometric, 244
        proportional, 247-248
    measurement tools, 240
        3D scanners, 242
        calipers, 241
    measuring parts, 243
    modeling parts, 248-249

Reverse Engineering project, 275-276
    adding screw hole, 281-282
    creating new part, 276
    creating screw hole support, 280-281
    cutting socket opening, 278-279
    exporting .STL file, 282
    extruding body, 276
    filleting edges, 276-277
    mirroring socket opening, 279
    printing, 282
    shelling cover, 277-278
revolve command (CAD), 219
revolved cut command (CAD), 219
rigidity, 3D printer frame, 54
Robot project, 193
    loading filament, 194
    loading .STL file, 193
    modifying slicer settings, 194-196
    monitoring first layer, 197
    preheating extruder and heated bed, 194
    printing feet down, 193
    removing, 198-199
    removing supports, 200-202
    slicing model, 196
    starting print, 197
rods
    smooth, 61-62
    threaded, 69-70
roller bearings, 62-63
rotational bearings, 63
running prints, 158-159

S

scaling, CAD (computer-aided design), 225-226
scanners (3D), 242
screws, lead, 68-69
SD (secure digital) cards, 109
    control board support, 110
selective laser melting (SLM), 42
selective laser sintering (SLS), 9, 15, 42

settings (slice software), 149
    filament, 152-154
        Bridges fan speed, 154
        Diameter, 153
        Disable fan for the first, 154
        Enable auto cooling, 153
        Enable fan if layer print time is below, 154
        Extrusion multiplier, 153
        Fan speed, 154
        First layer bed temperature, 153
        First layer extruder temperature, 153
        Keep fan always on, 153
        Min print speed, 154
        Other layers bed temperature, 153
        Other layers extruder temperature, 153
        Slow down if layer print time is below, 154
    print, 154-157
        Brim width, 156
        Distance from object, 156
        Fill density, 156
        Fill pattern, 156
        First layer height, 155
        Layer height, 155
        Minimum extrusion length, 156
        Multiple extruders, 156
        Perimeters (minimum), 155
        Raft, 156
        Skirt height, 156
        Skirt loops, 156
        Solid layers (bottom), 155
        Solid layers (top), 155
        Speed, 156
        Spiral vase, 155
        Support settings, 156
        Top/bottom fill pattern, 156
    printer, 150-152
        Bed size, 150
        End G-code, 151
        Extra length on restart, 152
        Extruders, 151
        G-code flavor, 150

Layer change G-code, 151
Lift Z, 151
Nozzle diameter, 151
Minimum travel after retraction, 152
Position, 151
Print center, 150
Retraction length, 151
Retract on layer change, 152
Speed, 152
Start G-code, 151
Tool change G-code, 151
Use firmware retraction, 151
Use relative E distances, 151
Vibration Limit, 151
Wipe while retracting, 152
shell command (CAD), 221
shields, Arduino control boards, 106-107
shrouds, fan, adding, 296-297
silicone heating pads, 93
sketching commands (CAD), 222-223
Skirt height setting (slicing software), 156
Skirt loops setting (slicing software), 156
SLA (stereolithography), 14-15
SLA (stereolithography) printers, 38
sleeve bearings, 63
Slic3r, 24-25
Slice with Slic3r button, 157
slicing layers for models, 24-26
slicing models, 177-178
slicing software, 24-25, 28, 131
    configuring, 148-157
        settings, 149
            filament, 152-154
            print, 154-157
            printer, 150-152
    versus host software, 131
SLM (selective laser melting), 42
Slow down if layer print time is below setting (slicing software), 154
SLS (selective laser sintering), 9, 15, 42
smooth rods, 61-62

software
    CAD (computer-aided design), 213, 227
        choosing, 217
        commands, 218-223
        exporting files, 236-237
        modeling, 228-236
        origins, 214
        premodeling, 228
        scaling, 225-226
        simulation add-ons, 214
        units, 223-224
        versus 3D modeling, 215-217
    firmware, 128-129
    host, 129-130
        connecting, 132-133
    slicing, 131
        configuring, 148-157
        versus host software, 131
solid bearings, 62-63
Solid layers (bottom) setting (slicing software), 155
Solid layers (top) setting (slicing software), 155
Solidworks CAD software, 217
specifications, 3D printers, 118
Speed settings setting (slicing software), 156
Speed (slicing software), 152
spindles, CNC mills, 27
Spiral vase setting (slicing software), 155
square-cube law, 119
standard tessellation language (STL), 14-15
Start G-code setting (slicing software), 151
Start Print button, 159
Star Trek, 10
stepper motors, 63-65
stereolithography (SLA) printers, 14-15, 38
STL (standard tessellation language), 14-15
stopping motor power command, 134
Storage Box project, 203
    loading drawer .STL file, 207-208
    loading storage body .STL file, 204-205
    printing drawer, 208-210
    printing storage body, 206-207
    slicing storage body, 205-206

Stratasys, 15, 94
stringing, troubleshooting, 171
subtractive manufacturing, CNC mills, 28-29
Support settings setting (slicing software), 156
surface finish, 11-12
surface finishes, CNC mills versus 3D printers, 32-34
surface treatments, build platform, 95-99
    ABS juice, 98
    hairspray, 98-99
    painter's tape, 95-96
    PET (polyethylene terephthalate), 97
    polyimide film, 96-97
    white glue, 96
sweep command (CAD), 219
swept cut command (CAD), 219
switches, end stops, 102-104

T

tangent lines, 220
temperature, printing filament materials, 291-292
temperature problems, troubleshooting, 166
    cracking, 168
    discoloration, 169
    hot ends, 166-168
text command (CAD), 222
thermistors, 82-83
thermocouples, 82-83
Thingiverse, 305
threaded rods, 69-70
3D modeling
    CAD (computer-aided design), 228-230
        assemblies, 231-233
        exporting files, 236-237
        fitting parts, 231-235
        modeling parts, 235-236
        premodeling, 228
    versus CAD (computer-aided design), 215-217
3D printers, 113
    assembled, 116-117
    attachment and connection components, 70-73

building DIY, 117-118
build platform, 91
    heated beds, 92-94
    heated build chambers/enclosures, 94
    materials, 92
    surface treatments, 95-99
Cartesian, 50, 122
    construction, 50-51
    layout, 51-53
choosing, 118-123
closed-source hardware, 116
control components, 101
    control boards, 106-109
    end stops, 102-105
    LCD controllers, 110-112
    SD cards, 109-110
controlling, 133-134
Delta-style, 122-123
DLP (digital light processing), 38-39, 123
evolution, 18-19
extruders, 31, 75
    cold ends, 76-81
    hot ends, 81-85
    multiple, 87-88
    nozzles, 85-86
    print fans, 86-87
FFF (fused filament fabrication), 42-45
fire hazards, 11
frames, 49
    Cartesian layout, 50-53
    construction, 53-56
    size, 56-57
hot ends, 26
kits, 117
manufacturing, additive, 24-27
modifying, 295
    all-metal hot ends, 300
    bed size, 297-299
    control boards, 298
    converting to PCB mills, 302-303
    extending axes, 302
    fan shrouds, 296-297

finding parts, 304-305
heated beds, 297
installing multiple extruders, 301-302
laser cutting, 303-304
power supplies, 298-299
movement components, 59-60
bearings, 62-63
belts, 65-66
pulley systems, 65-66
rails, 61-62
smooth rods, 61-62
stepper motors, 63-65
open source, 114-116
overhangs, 11
parts, availability, 20-21
powder bed printing, 39-40
prices, 21-22, 121-122
print resolution, 119-120
print volume, 119
rewiring, 298
SLA (stereolithography), 38
SLS (selective laser sintering), 42
software
firmware, 128-129
host, 129-133
slicing, 131, 148-157
specifications, 118
unusual designs, 123
versus CNC mills, 23-24, 29, 34
cost, 29-30
materials, 32
part geometry, 30-31
surface finishes, 32-34
weight-bearing components, 67
lead screws, 68-69
threaded rods, 69-70
3D printing
benefits to business, 4-7
benefits to hobbyists, 7-9
finishing, 11-12
misconceptions, 9-11
origins, 13-16

3D scanners, 242
3D Systems, 15
Tinkercad, 217
Tool change G-code setting (slicing software), 151
Top/bottom fill pattern setting (slicing software), 156
trim command (CAD), 223
troubleshooting prints, 161-162, 171-173
adhesion problems, 169-171
extrusion problems, 163-166
temperature problems, 166-169
t-slot aluminum extrusion, printer frames, 55-56

U-V

underextrusion, troubleshooting, 164
units, CAD (computer-aided design), 223-224
upgrading control boards, 108-109
USB cable, control board, connecting to computer, 132
Use firmware retraction setting (slicing software), 151
Use relative E distances setting (slicing software), 151

vertical surfaces, layer ridges, 11-12
Vibration Limit setting (slicing software), 151

W-X-Y-Z

warping
FFF printers, 44-45
troubleshooting, 170
weight-bearing components, 67
lead screws, 68-69
threaded rods, 69-70
white glue, build platform surface treatment, 96
Wipe while retracting setting (slicing software), 152
wood filament, 287

Z Corp., 15
Z height, 139
adjusting, 144-146
knowing correct, 145-146
leveling, auto-leveling systems, 171